# Studies
# in the History of Mathematics and Physical Sciences

## 6

John T. Cannon
Sigalia Dostrovsky

# The Evolution of Dynamics: Vibration Theory from 1687 to 1742

With 10 Illustrations

Springer-Verlag
New York   Heidelberg   Berlin

JOHN T. CANNON
and
SIGALIA DOSTROVSKY
155 Fairfield Pike
Yellow Springs, Ohio 45387/USA

AMS Subject Classifications: 01A45, 01A50, 73-03, 73 D30

**Library of Congress Cataloging in Publication Data**

Cannon, John T.
    The evolution of dynamics: vibration theory from 1687 to 1742.
    (Studies in the history of mathematics and
physical sciences; 6)
    Bibliography: p.
    Includes index.
    1. Vibration—History—17th century.
2. Vibration—History—18th century.
I. Dostrovsky, Sigalia.  II. Title.
III. Series.
QA865.C36      531'.32      81-14353
ISBN 0-387-90626-6          AACR2

Printed in the United States of America.

9 8 7 6 5 4 3 2 1

ISBN 0-387-90626-6   Springer-Verlag New York Heidelberg Berlin
ISBN 3-540-90626-6   Springer-Verlag Berlin Heidelberg New York

7-6-91

# Preface

In this study we are concerned with Vibration Theory and the Problem of Dynamics during the half century that followed the publication of Newton's *Principia*. The relationship that existed between these subjects is obscured in retrospection for it is now almost impossible not to view (linear) Vibration Theory as linearized Dynamics. But during the half century in question a theory of Dynamics did not exist; while Vibration Theory comprised a good deal of acoustical information, posed definite problems and obtained specific results. In fact, it was through problems posed by Vibration Theory that a general theory of Dynamics was motivated and discovered.

Believing that the emergence of Dynamics is a critically important link in the history of mathematical science, we present this study with the primary goal of providing a guide to the relevant works in the aforementioned period. We try above all to make the contents of the works readily accessible and we try to make clear the historical connections among many of the pertinent ideas, especially those pertaining to Dynamics in many degrees of freedom. But along the way we discuss other ideas on emerging subjects such as Calculus, Linear Analysis, Differential Equations, Special Functions, and Elasticity Theory, with which Vibration Theory is deeply interwound. Many of these ideas are elementary but they appear in a surprising context: For example the eigenvalue problem does not arise in the context of special solutions to linear problems—it appears as a condition for isochronous vibrations.

Although mathematical thought differs in different ages, mathematics itself has a coherence that transcends time. Thus it provides a powerful tool with which to grasp modes of thought from former times. From an immersion in the details of mathematical arguments, one can gather enough precise understanding to be able to enter into the domain of the intuitive. Therefore we believe that our study not only describes a link in the evolution of a specific subject but also that it assists in the attainment of a feel for physics in the age of Newton and the Bernoullis.

In spite of its evident importance, dynamics in the first half of the eighteenth century has been largely neglected. This is the period of late Newton and early Euler; thus it lies in the shadow of great brilliance coming from both before and after. For example, Euler was a central figure; but his works from the period go with little notice because he later reworked everything in a form and from a point of view that have become generally familiar. Thus Truesdell's notes on Euler were pioneering works.[1] Truesdell emphasized the fact that the idea of dynamical equations was slow to emerge; furthermore, he provided a basic indication of the contents of a vast number of papers, including most of the papers considered in the present study. We gratefully acknowledge our indebtedness to his notes.

Yellow Springs                                                    J.C. and S.D.
April 1981

---

[1] Truesdell [1, 2].

# Table of Contents

# 1. Introduction

Before the middle of the eighteenth century, no-one had any notion that "Newton's Second Law" could be used as the basis for a dynamical description of a mechanical system that has several degrees of freedom. Yet by name and by tradition this notion is associated with the *Principia* of 1687 and it is commonly supposed that Newton, at least, knew that his "Second Law" described in principle the motions of a mechanical system. What is true, is that a lot was understood before the middle of the eighteenth century about the dynamics of systems in one degree of freedom (as well as about a few systems having special symmetries in a few degrees of freedom) and Newton was indeed the greatest master of the period. Thus the common supposition about Newton's understanding of dynamics is created by ignoring the enormous distinction that separates mechanical systems in one degree of freedom from those in more than one or, as we will say, *many degrees of freedom*. On a popular level, this distinction has the sound of a mere technicality, concerned only with the details of complexity, and it may be for this reason that it has been so commonly ignored. One might wish for an adjective of moment to describe the difference encountered when there are many degrees of freedom. (One can talk about the *dimension* of the manifold of configurations; but in the present work, dimension will refer only to physical space.) The purpose of the present study is to follow in detail those works that dealt with the dynamics of systems in many degrees of freedom up to the year 1742 (apart from the cases, like the central force problem, that were handled because of their special symmetries).

Thus, to grasp how it was that "Newton's Second Law" had a different appearance before the middle of the eighteenth century, one should distinguish between systems in one and in many degrees of freedom. We will incorporate in our terminology a second distinction of a different type which refers to a conceptual limitation. "Newton's Second Law" in its modern sense will be referred to as the *momentum principle*. Thus the momentum principle (together with the corresponding principle for angular momentum) entails the idea that a system of equations determines a system's motion for given initial conditions. The equations require that one specify forces even for states that are never realized in particular motions of interest. But "Newton's Second Law," as it was understood before the middle of the eighteenth century, will be referred to as the *momentum law*. The momentum law, then, is to be understood as a condition on a particular motion of a mechanical system: the mass of an element of the system multiplied by its acceleration must be equal to the force which that element experiences during a given motion of the entire system. This is a condition on a given motion and it is not a system of dynamical equations. In practice, the *momentum law* is a consistency condition that typically leads to a constant of a particular motion such as the period of an oscillation. We will see this in examples, beginning with Newton's own work which we will discuss in Chapter 2.

In the early eighteenth century, functional notation was used only in very limited circumstances. In recognizing that certain concepts have a vastly different appearance when a formulation from the point of view of function theory is not available, one obtains a third distinction that helps to clarify the early eighteenth century perspective on dynamics. For example, if the momentum law had been formulated functionally, the force on each element would have been given as a function of the state of the system, and the momentum principle would have been quickly noticed. That is, the difference between the momentum law and the momentum principle can be maintained only as long as there is no general freedom or incentive to make functional formulations.

In the case of a single degree of freedom, there was not a significant limitation to giving a functional formulation of the momentum law and, in fact, there was, in this case, occasional use of the momentum principle in the early eighteenth century. But this use was not sufficient to suggest seriously that a momentum principle could be used as a basis for dynamics. Actually, the momentum principle played only an auxiliary role even for the cases of one degree of freedom since, for a conservative system, it was written in the form $mv\,dv = F\,dy$ and it was immediately integrated to obtain what we would call conservation of energy. Since the most typical problem of this type concerns a particle constrained to a curve in the gravitational field, conservation of energy itself was the most natural starting point and it sufficed for dynamics.

The question of functional formulations arose also in connection with techniques of calculus. In fact, during the eighteenth century, *geometrical calculus* gradually gave way to *functional calculus*. The distinction between the two calculi is clear in the following example: the equation $dy = \beta\sqrt{(c^2 - y^2)}\,dt$ can be understood geometrically through the construction in Figure 1.1 with its implied limiting situation. It is understood functionally if one sets $y = c\sin(\beta t + \delta)$ with the sine understood as a function. Taylor (Chapter 3) and Johann Bernoulli (Chapter 8) discovered that in its funda-mental mode, a vibrating string has the shape of a sine function—though they did not express it this way. They expressed it geometrically and they did not discover that also in the higher modes the string has the shape of a sine function. That is, from the functional point of view, one naturally

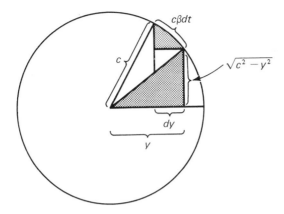

Figure 1.1

considers the sine to be defined on the whole real line while, from the geometrical point of view, angles larger than $2\pi$ are not very natural.

However, as regards techniques of calculus, we will use functional notation because it is familiar and more concise. But we will try to keep the reader reminded of the situation when the relevant calculus is geometrical. Also, we will make references such as "the integration of $dy = \beta\sqrt{(c^2-y^2)}\,dt$ was familiar geometry." There is, admittedly, something misleading in describing early eighteenth century work with the help of functional notation. On the other hand, in following an earlier work, one does have to comprehend it as thoroughly as would a contemporary reader (not to mention the fact that one also wants to read with historical perspective). A contemporary reader would have been imbued with contemporary mathematical techniques that are obsolete and naturally forgotten today. One can make a study of these things; but one can also view them as a distraction when pursuing something else. In any event, what we offer is a guide to the works, not a substitute.

\*       \*       \*

When one looks for problems in dynamics in many degrees of freedom that were treated before the middle of the eighteenth century, one finds problems in vibration theory. Since these are linear and solvable problems in dynamics, it may not be surprising that they were the first to be considered. But it has to be emphasized that there was no theory of dynamics and therefore no notion of what a linear problem in dynamics might be. From a twentieth century point of view one should also expect the linear problems in two degrees of freedom to have been considered first—after the case of one degree of freedom. But they were not. In fact, the cases of a few degrees of freedom were considered rather incidentally, while the guiding problems involved infinitely many degrees of freedom. These were, for example, the problems of the propagating pressure wave and the vibrating string. They were chosen not because they were simple examples in a general theory of dynamics but because they came from a subject that already had its own identity.

Thus, to see the emergence of dynamics, we must turn to vibration theory at a time when that theory was not guided by any qualitative understanding of dynamics in many degrees of freedom. At the end of the seventeenth century, vibration theory already had a strong tradition.[1] In a sense, partly mythological, this tradition went all the way back to Pythagoras who had quantified pitch with string length. But it was a seventeenth century discovery that pitch was appropriately quantified by frequency; so the Pythagorean arithmetic of Medieval music theory gave a complete experimental classification of relative frequencies. The seventeenth century provided Mersenne's law, that the frequency $\nu$ of a vibrating string of length $l$, tension $P$, and linear mass density $\rho$, satisfies the proportionality

$$\nu \propto \frac{1}{l} \sqrt{P/\rho}. \tag{1.1}$$

Consciously entailed in this is the notion of *isochronism*, namely frequency's independence of amplitude. Along with Mersenne's law appeared 1) the problem of deriving it, 2) the problem of deriving the constant of proportionality, and 3) the problem of measuring absolute frequency. One had the problem of doing for the string what Huygens had done for the pendulum. But lest one should solve these problems, there was more. The seventeenth century also discovered, demonstrated and described the higher vibrational modes. It perceived an analogy between the vibrating string and the vibrations of air in a pipe. It provided qualitative descriptions of sound in air, some quite good and others wrong. It presented the problem of finding the velocity of sound and observed its independence of pitch and volume. It gave discussions on consonance, dissonance, superposition, and beats. It left voluminous numbers of observations on special phenomena—always with an accompanying problem. How can one use parameters to determine the pitch of a bell? What are a bell's best dimensions? Beside this tradition, the double pendulum had but modest appeal. It was the problem of sound velocity and the derivation of Mersenne's law that were treated first and the intuitions of vibration theory guided the formulation of concepts.

The now obvious approach to the theory of small vibrations is to linearize the dynamical equations at a stable configuration. Not only were dynamical equations unavailable in the period under consideration but there was no

---

[1] The development of vibration theory in the seventeenth century is discussed in Dostrovsky [1] and Dostrovsky and Cannon [1].

systematic notion of linear analysis nor of the associated concepts such as eigenvalues, eigenvectors, sinusoidal solutions, modes, superposition, and so on. In fact, vibration theory in the first part of the eighteenth century is a prelude to the discovery of linear analysis. On the other hand, the concepts that were prevalent in this epoch of the evolution of vibration theory now seem a little diffuse though they were phenomenologically relevant. In addition to isochronism, there was the notion of *simultaneous crossing of the axis*. This refers to the fact that the systems considered—the taut string, the hanging chain, the straight rod, and their discrete analogues—having equilibrium configurations along axes were observed to undergo simpler motions when the motions took them through their equilibrium configurations, that is, when all parts of a system reached its axis simultaneously. Thus, the motions that were studied were those having the properties of isochronism and simultaneous crossing of the axis. The connection between isochronism and the simple pendulum was familiar in the work of Beeckman, Galileo, and Huygens. It was clearly assumed that isochronism meant the restriction to the case of small vibrations. But furthermore, it was assumed that isochronism and simultaneous crossing of the axis meant that each element of the system would move as a simple pendulum, that is, each element would undergo simple harmonic motions having the same period. This assumption we will refer to as the *pendulum condition* and it will save us repeated fuss if we introduce once and for all a notation to handle it. If $F$ is the force on the mass element $m$ when it is displaced from the axis (its equilibrium position) by $y$, then the pendulum condition is

$$aF = -my \qquad (1.2)$$

for all mass elements of the system. (If the mass element is infinitesimal, $m = \rho \, dV$, then the force is also, $F = f \, dV$ and one writes (1.2) with the $dV$'s cancelled.) We will use the letter $a$ systematically as in equation (1.2). Thus, since $F = m\ddot{y}$, (1.2) gives the equation of motion $a\ddot{y} = -y$ which has the solutions $y = c \sin(t/\sqrt{a} + \delta)$ and hence the period and frequency

$$T = 2\pi\sqrt{a} \quad \text{and} \quad \nu = \frac{1}{2\pi\sqrt{a}}. \qquad (1.3)$$

In the case of a simple pendulum of length $l$ undergoing small oscillations, $F = (-gm/l)y$. That is,

$$l = ag, \qquad (1.4)$$

which will be referred to as the *length of the simple isochronous pendulum*, meaning that it is the length of a pendulum having the same period as does the motion under consideration. Typically the authors of the period look for $\mathfrak{a}g$ rather than $T$ or $\nu$. We will refer to $\mathfrak{a}$ as the *harmonic constant* and to $\mathfrak{a}^{-1}$ as the *intensity* of the *harmonic force* $-\mathfrak{a}^{-1}my$. Note that

$$\mathfrak{a}v_{max}^2 = y_{max}^2, \quad \mathfrak{a}F_{max} = my_{max}, \quad \text{and} \quad \mathfrak{a} = \frac{m^2 v_{max}^2}{F_{max}^2} \qquad (1.5)$$

where $v = \dot{y}$. When it happens that the mass and various constant factors that appear in the force are most conveniently incorporated in the harmonic constant, we will warn the reader by writing $\hat{\mathfrak{a}}$. For example, the vibrating rod has the *pendulum condition* $\hat{\mathfrak{a}}\, d^4y/dz^4 = y$ where $\hat{\mathfrak{a}} = \mathfrak{a}e/\rho$, $\rho$ being the linear mass density and $e$, the flexural rigidity.

But to reiterate the historical statement made above: It was believed throughout the period before the middle of the eighteenth century that isochronism and simultaneous crossing of the axis always imply the pendulum condition. It was a shocking surprise when d'Alembert showed that the vibrating string itself provides a counter example. Motions not having the properties of isochronism and simultaneous crossing of the axis were called *irregular* and they were not treated.

*       *       *

It is often remarked that Newton used Boyle's law to derive the speed of sound and, of course, that an adiabatic correction is needed. However, here (Chapter 2), we approach Newton's analysis of the pressure wave with the primary question: how in fact did Newton deal with a dynamics problem in many degrees of freedom? In scrutinizing the details of Newton's analysis, one comes to face the ultimate limitations of seventeenth century mechanics. Newton used the *momentum law* but he clearly did not have the *momentum principle* at his disposal; he gave verbal descriptions and geometrical arguments, not having functional notations or concepts to handle these things. Once one is acclimatized to the limitations, one appreciates Newton's mastery. Though Newton's reputation is untarnishable, his analysis of the pressure wave has been treated rather casually in the recent literature. Laplace, however, wrote that "Newton's theory, although imperfect, is a monument to his genius."[2]

Taylor followed Newton in his approach to dynamics in many degrees of freedom (Chapter 3). But Taylor had a background in music and it was not unnatural that he should take up the problem of the vibrating string, as he did in 1712. He introduced the main ideas that dominated the mathematical approach to vibration theory in the early eighteenth century: *isochronism* and *simultaneous crossing of the axis* and their supposed implication that the *pendulum condition* must be satisfied. Because he was the first to derive Mersenne's law (1.1) with its constant of proportionality,

[2] Laplace [1], V, p. 95.

Taylor is rightfully famous. But his presentation was concise to the point of being cryptic and remarks in the recent literature have indicated either what he should have done or else the confusion which his presentation fails to disallow.

It took about twenty years for Taylor's analysis of the vibrating string to be assimilated. In reading papers that were written during this period one learns a good deal about the climate of opinion concerning dynamics in many degrees of freedom. For example, Hermann, who was expert in the dynamics of one degree of freedom, published comments on both Newton's and Taylor's works. He generalized the momentum principle to the case of many degrees of freedom in a completely fallacious way (Chapter 5). Euler, who would later be the proper discoverer of the momentum principle, first attempted to treat the dynamics of the vibrating ring by generalizing energy conservation (which suffices for one degree of freedom) in effect to a principle of local conservation of energy (Chapter 7). In 1728, Johann Bernoulli, who earlier had claimed that Taylor had done nothing but plagiarize, did take a serious look at Taylor's analysis and began analyzing the loaded string by the same method (Chapter 8). At the same time that Taylor did his work, Sauveur also derived Mersenne's law with its constant. Sauveur treated the vibrating string as a compound pendulum and thus he was able to avoid the problem of many degrees of freedom (Chapter 4).

However, around 1733 Daniel Bernoulli and Euler dealt with the hanging chain. They worked easily with the methods of Taylor. They consciously sought "Dynamical Principles," but on finding forces they used them only for the *pendulum condition*. They freely used the functional point of view as far as one variable is concerned and thus they discovered Bessel functions, Laguerre polynomials, some of their zeros and the higher modes of vibration. In the geometric point of view, boundary value problems had been camouflaged; but Bernoulli recognized the generality accomplished in treating them analytically. Thus he showed, for example, how to treat the case of a chain hung from a mass that is suspended by a string. Meanwhile, Euler obtained the Poisson–Lommel integral representation of Bessel functions (Chapter 9). (In the appendix we have included facsimiles and translations of Bernoulli's papers on the hanging chain. These papers can be considered as central to our subject, not only in importance but also as a representation of style. Like most of his papers, these are forthright in showing his intent. They show the balance between geometrical and functional calculus that is typical of the last decade of our period of study.)

Following the approach they had used to treat the oscillations of the hanging chain, Daniel Bernoulli and Euler went on to treat other oscillations. In their work, the *pendulum condition*, which of course survives as an eigenvalue problem, became standardized as a basic equation, though their ideas on dynamics itself are forgotten. Their results on the vibrating rod (or beam) are classic. With this problem they learned to use boundary conditions to specify solutions of higher order differential equations (though they had more difficulty in obtaining the actual solutions than one might expect). In aural experiments with the vibrating rod, Bernoulli observed superpositions and pointed out their theoretical possibility (Chapter 13). Euler began to look on the pendulum condition as a condition of static equilibrium, thereby putting a dynamical problem into a form that would be naturally treated by variational methods (Chapter 10).

The other problems that they treated concentrated on cases with only a few degrees of freedom: namely the rocking and bobbing of a floating body (Chapter 12) and the linked compound pendulum (Chapter 14). The rocking of a symmetrical floating body was of practical concern for ship design and notions such as that of the "metacenter" of a ship were introduced at this time. But the asymmetric case where bobbing and rocking both occur in a simple mode was more instructive for mechanics.

Finally in 1742, at the very end of the epoch under consideration, Johann Bernoulli published his own versions of the oscillations of a linked pendulum (Chapter 15). In the case of two links, he first found equations that applied to motions that need not be small. To do this, Bernoulli used the momentum law. But, characteristically of the epoch, Bernoulli did not attempt to find enough equations to determine the motion except in the limit of small oscillations. In this limit, Bernoulli found equations that were a great deal more complicated than those that result from the pendulum condition. Bernoulli's experience would not have encouraged the idea that it is practicable to find dynamical equations to describe motion, much less that the momentum law would suffice for this. Bernoulli proceeded to give another method for treating the small oscillations, namely the standard method of the pendulum condition and this he found much more natural.

Bernoulli was an elder statesman of mathematical science at the close of our epoch. His work serves well to characterize the contemporary perspective on dynamics in the half-century after Newton's *Principia*.

# 2. Newton (1687)

Newton's analysis of the propagating pressure wave (applicable both to sound and to light) appeared in his *Principia* of 1687 and, with an unfortunate change in the order of presentation, in the later editions of 1713 and 1726. It is here that we find the first use of the *momentum law* applied to an element of a continuous body. We will begin with a detailed discussion of *Prop.* XLVII of Book II (which in the second and third editions becomes *Prop.* XLVIII), following Motte's translation[1] of 1729. In section 2.2 we will remark on Cotes' introduction of the change in the order of presentation and on some of the confusions that resulted.

## 2.1. Pressure Wave

Newton's *Prop.* XLVII reads:

The velocities of pulses propagated in an elastic fluid, are in a ratio compounded of the subduplicate ratio of the elastic force directly, and the subduplicate ratio of the density inversely; supposing the elastic force of the fluid to be proportional to its condensation.

Now, the demonstration is entirely verbal and we shall go through the exercise of translating it; in preparation for this, let $z$ label a point of the one dimensional continuum by its position in the equilibrium state and let its position at time $t$ be written as $z + y(z, t)$. Newton will be describing a periodic wave traveling in one direction and he will use the relation between position and time that allows us to write $y(z, t) = Y(t - (z/V))$, where $V$ is the velocity of propagation, with $Y$ periodic, $Y((l - z)/V) = Y(-z/V)$, so that $l$ is the wave length. Let $\rho$ denote the density of the continuum in its equilibrium state. Under the displacement $y$, the element which in equilibrium is $\rho\,dz$ goes into the element $\rho(1 + \partial y/\partial z)\,dz$, to lowest order in $\partial y/\partial z$. The relative extension or strain of this element, or, in Newton's

---

[1] Newton [1].

words, its "dilatation or contraction" is

$$\mu(z, t) = \frac{\partial y}{\partial z}. \tag{2.1}$$

By Boyle's law, the excess pressure in the element is $-P\mu$ where $P$ is the equilibrium pressure and, as Newton will say, the force on the element is then

$$P \, d\mu = P \frac{\partial \mu}{\partial z} \, dz. \tag{2.2}$$

Now, if one has a knowledge of partial differentiation and of the *momentum principle*, one simply puts (2.1) and (2.2) together to obtain the wave equation,

$$\rho \frac{\partial^2 y}{\partial t^2} = P \frac{\partial^2 y}{\partial z^2}. \tag{2.3}$$

and one sees that $Y(t-(z/V))$ is a solution if $V = \pm\sqrt{P/\rho}$. Without this however, Newton will consider the periodic case and, holding $l$ fixed, he will scale $y$ to $\gamma y$; whence he will have, as we do by (2.1), that the strain $\mu$ scales to $\gamma\mu$ or that the excess pressure $-P\mu$ scales to $-\gamma P\mu$ or finally that the force $P \, d\mu$ scales to $\gamma P \, d\mu$. Next, by the momentum law (and not the momentum principle!) he will have that the velocity $\partial y/\partial t$ scales to $\gamma(\partial y/\partial t)$; hence he will have found that the velocity scales with the amplitude so that the period doesn't change. Finally, since $l$ and the period remain the same under the scaling, he will point out that the scaled waves advance at the same speed as do the unscaled, namely $V$. Now we turn to Newton's version with the warning that statements of the form "$A$ is as $B$" must be read as "$A$ scales as $B$" and not as "$A$ is proportional to $B$ at all times."

### Case 1

If the mediums be homogeneous, and the distances of the pulses [wavelengths] in those mediums be equal amongst themselves, but the motion [$y$] in one medium is more intense than in the other: the contractions and dilatations [strain] of the correspondent parts will be as those motions. Not that this proportion is perfectly accurate. However, if the contractions and dilatations are not exceedingly intense, the error will not be sensible; and therefore this proportion may be considered as physically exact. Now the motive elastic forces are as the contractions and dilatations; and the velocities generated in the same time in equal parts are as the forces. Therefore equal and corresponding parts of corresponding pulses will go and return together, through spaces proportional to their contractions and dilatations, with velocities that are as those spaces; and therefore the pulses, which in the time of one going and returning advance forwards a space equal to their breadth, and are always succeeding into the places of the pulses that immediately go before them, will, by reason of the equality of the distances, go forward in both mediums with equal velocity.

So far, Newton has found that $V$ is independent of amplitude; next he will find that it is independent of the wave length $l$. This time he will scale $z$ to $\gamma z$ and $l$ to $\gamma l$ and, since by the preceding part he is free to scale $y$, he will scale $y$ to $\gamma y$. Then by (2.1) the strain will be unchanged; and therefore he will find that the force $P(\partial \mu / \partial z)\, dz$ remains unchanged under the scaling. Next, by (2.2) and the *momentum law* he will have that the acceleration $\partial^2 y / \partial t^2$ scales to $(1/\gamma)(\partial^2 y / \partial t^2)$ and so $t$ must scale to $\gamma t$ and the period $T$ must scale to $\gamma T$. Thus, since $l$ and $T$ scale together, he will find that the wave velocity $V$ remains unchanged.

### Case 2

If the distances of the pulses or their lengths are greater in one medium than in another; let us suppose that the correspondent parts describe spaces $[y(z, t)]$, in going and returning, each time proportional to the breadths of the pulses: then will their contractions and dilatations be equal. And therefore if the mediums are homogeneous, the motive elastic forces, which agitate them with a reciprocal motion, will be equal also. Now the matter to be moved $[\rho\, dz]$ by these forces is as the breadth of the pulses; and the space through which they move every time they go and return, is in the same ratio. And moreover, the time of one going and returning, is in a ratio compounded of the subduplicate ratio of the matter, and the subduplicate ratio of the space; and therefore is as the space. But the pulses advance a space equal to their breadths in the times of going once and returning once, that is, they go over spaces proportional to the times; and therefore are equally swift.

At this point, Newton has found that $V$ is independent of $l$ as well as of the amplitude; next he will find its dependence on $\rho$ and $P$. From (2.2) and the *momentum law*, he will conclude that, if $\rho / P$ is scaled to $\gamma \rho / P$ while $z$ and $y$ remain the same, the time $t$ will be scaled to $\sqrt{\gamma}\, t$ which means that the velocity of propagation will scale to $(1/\sqrt{\gamma})V$. That is, that $V \propto \sqrt{(P/\rho)}$.

### Case 3

And therefore in mediums of equal density and elastic force, all the pulses are equally swift. Now if the density or the elastic force of the medium were augmented, then because the motive force is increased in the ratio of the elastic force, and the matter to be moved is increased in the ratio of the density; the time which is necessary for producing the same motion as before, will be increased in the subduplicate ratio of the density, and will be diminished in the subduplicate ratio of the elastic force. And therefore the velocity of the pulses will be in a ratio compounded of the subduplicate ratio of the density of the medium inversely, and the subduplicate ratio of the elastic force directly.

This completes the demonstration. Thus, Newton assumes something about the motion, namely in verbal form that $y = Y(t - (z/V))$; from this he finds from the momentum law that $V \propto \sqrt{(P/\rho)}$. He does not use the momentum principle to determine the motion.

In the next two propositions, *Prop.* XLVIII (which in the second and third editions becomes *Prop.* XLVII) and *Prop.* XLIX, Newton takes an explicit example for what we have designated in functional notation by $Y$, namely $Y(t) = r[1 - \cos \omega t]$ where $\omega = 2\pi V/l$. Before we indicate his method of calculation, we can give the following idealized outline: the force (2.2) takes the maximum value $P(4\pi^2/l^2)r\,dz$; it acts on the mass $\rho\,dz$ which undergoes simple harmonic motion with amplitude $r$. Hence, by (1.5) and (1.3), the motion has the period $T = l\sqrt{(\rho/P)}$. Finally, the wave velocity is $V = l/T$. That is,

$$V = \sqrt{P/\rho}. \tag{2.4}$$

Newton's discussion of this is colored by the fact that he immediately views a velocity in the form $V = \sqrt{\mathfrak{g}A}$ not only as the velocity that a freely falling mass acquires in descending from a height $A/2$ but also as that velocity which will take an object around a circle of radius $A$ in the time of one oscillation of a simple pendulum of length $A$. More significantly, Newton does not use the strain (2.1) in calculating the force (2.2); he calculates the force using the differential of the total pressure $P(1+\mu)^{-1}$. Thus, he does not illustrate the concept of strain which he has grasped in the preceding proposition. The main details of Newton's calculation, which is of course geometrical, can be indicated as follows: $\tau = t - (z/V)$ and $d\tau = -dz/V$ are given as a point and an infinitesimal arc on a circle. $Y(\tau) = r[1 - \cos \omega\tau]$ is given as the projection to a diameter and $dY/\omega d\tau = r \sin \omega\tau$, as the perpendicular. Newton calculates the ratio[2] $dz/(1+\mu)\,dz = V/(V - \omega r \sin \omega\tau)$. By Boyle's law, Newton has that the pressure at $z$ at time $t$ is $P$ times this ratio. He calculates the force on the element $\rho\,dz$ as $P$ times the difference between this ratio and the one evaluated at $\tau + d\tau$. In *Prop.* XLVIII, using the fact that $r$ is small, he simplifies the difference to find the force[3] $P(4\pi^2/l^2)r \cos \omega\tau\,dz$; in *Prop.* XLIX he calculates the extreme value.

It seems clear to us that Newton's purpose in *Props.* XLVIII and XLIX is to carry out these calculations and arrive at the result (2.4) for the velocity of sound. However, *Prop.* XLVIII is an unfortunate statement that has caused a great deal of confusion. It asserts that small amplitude pressure waves are always simple harmonic. But in the demonstration he says, "Let us suppose, then, that a medium hath such a motion excited in it from any cause whatsoever, and consider what will follow from thence;" and in the

---

[2] In Newton's terminology and notation: "the expansion of the part EG [$dz$], or of the physical point F in the place $\varepsilon\gamma$ is to the mean expansion of the same part in its first place EG, as V − IM is to V . . .", where E is our $z$, G is our $z + dz$, and F is a point in between, and $\varepsilon$ is our $z + y$, $\gamma$ is our $z + dz + y(z + dz)$, V is our $V/\omega$ and IM is our $r \sin \omega\tau$.

[3] In Newton's notation: (*Prop.* XLVIII) "the difference of the forces is to the mean elastic force of the medium [our $P$] . . . as HL − KN is to V," where HL is our $r \sin \omega\tau$, KN is our $r \sin \omega(\tau + d\tau)$, and V is our $V/\omega$; so (HL − KN)/V is our $(4\pi^2/l^2)r \cos \omega\tau\,dz$. Then in *Prop.* XLIX, Newton gives the extreme value HK/V, which is our $(4\pi^2/l^2)r\,dz$.

*Corollary* he says, "...the number of the pulses propagated is the same with the number of the vibrations of the tremulous body...". It appears to us that Newton thought 1) that a tremulous body undergoing small vibrations must have simple harmonic motion and 2) that the motion which it would excite in the medium must be harmonic also. Of course, 1) is false but understandable because the tremulous body in the idealized situation considered seems to have only one degree of freedom and 2) is correct: e.g. consider the solution of the wave equation on $z > 0$ with $y(z, t) = 0$ for $t \leq 0$ with the boundary condition $y(0, t) = 0$ for $t \leq 0$ or $t \geq 2\pi N/\omega$ and $y(0, t) = r[1 - \cos \omega t]$ for $0 \leq t \leq 2\pi N/\omega$; this is essentially the situation which Newton envisions in the Corollary.

## 2.2. Remarks

Newton's three propositions initiated dynamics in many degrees of freedom. Though influential, they were largely misunderstood, as we shall see in following later works. They were ingenious and also extremely obscure, especially in connection with the assumption of simple harmonic motion, and, above all, they gave a prediction for the velocity of sound that was wrong by 20%. The obscurity was much encouraged by the interchange of *Props.* XLVII and XLVIII, in the second and third editions, which put the general argument in the middle of the special argument that used simple harmonic motion. We will note some of Cotes' confusions which can be followed in his letters with Newton on the editing of the second edition.[4]

In his letter of 23 June 1711, Cotes writes to Newton, "I think ye $47^{th}$ Proposition is out of its place, for ye demonstration of it proceeds upon ye supposition of ye truth of ye $48^{th}$..." and he suggests the interchange. Newton accepts the interchange in his belated letter of 28 July in which he closes disinterestedly, "I will write to you about the third book in my next." It seems to us that Newton was reluctant to reconsider his argument; but he could have accepted the interchange on the grounds that it presented a concrete example before the general verbal argument. Were this the case he should have rewritten the presentation. Cotes was wrong but we will speculate on why he thought *Prop.* XLVII needed *Prop.* XLVIII: In *Prop.* XLVII Newton writes, as we have quoted above, that "the contractions and dilatations... will be as [the] motions" and that "the motive elastic forces are as the contractions and dilatations" and he is not very precise as to what exactly contractions and dilatations are. Putting these two parts together one finds that the motive elastic forces are "as" the motions. Taken out of context, there are two ways to read this: 1) the force is proportional to the displacement throughout the motion and 2) the force

---

[4] Newton [2], letters nos. 854, 857, 860, 863, 871, 889, and 891.

scales as the displacement. We hope it is clear from our presentation above that the second interpretation is correct. We interpreted contractions and dilatations as strain; but in any case, there is no conceivable interpretation which would make them proportional to displacement. Nevertheless, it seems that Cotes took the first interpretation which means that the motion must be simple harmonic.

The letters between Newton and Cotes provide some additional examples of the difficulties that Newton's contemporaries experienced in following the argument. In his letters of 23 June and 30 July, Cotes objects to *Prop.* XLVIII and its Corollary. In the notation we used above, Cotes' objection was that $z$ could not be both the position of a point of the continuum in its equilibrium state and the point's extremum position in a state of harmonic vibration. His reason was that it is at the mid-point of an harmonic vibration that there are no forces acting and hence only the mid-point could be a rest position.[5] In his letter of 28 July, Newton had tried to explain that the equilibrium position $z$ could appear at any point of the vibration, depending only on the boundary condition $y(0, t)$, and in the case $y(0, t) = r[1 - \cos \omega t]$ for $0 \le t \le 2\pi N/\omega$, mentioned above, that $z$ would be an extreme, as we may put it anachronistically. In his letter of 2 February, Newton finally responds with a physical description:

For the particles of air go from their *loca prima* with a motion accelerated till they come to the middle of the pulses where the motion is swiftest. Then the motion retards till the particles come to the further end of the pulses. And therefore the *loca prima* are in the beginning of the pulses. There the force is greatest for putting ye particle into motion if any new pulses follow.

In his letter of 7 February, Cotes accepts this; but there is no sense that he had come to understand it. Cotes' difficulty was that he applied the dynamics of one degree of freedom to the case of a particle in many degrees of freedom. This difficulty was typical. Somehow it seemed to a number of Newton's readers that the medium itself determines that the motion must be simple harmonic[6] and in fact of a given frequency (as though the medium were an harmonic oscillator in one degree of freedom). We shall see this with Hermann (Chapter 5) and Johann II Bernoulli (Chapter 11) who suggested that different frequencies involve different "fibers" corresponding to Mairan's suggestion that air is made up of different kinds of particles that transmit different frequencies (see 6.2).

---

[5] In their discussion of this point, the editors incorrectly accept Cotes' objection (*ibid.*, letter no. 854, p. 171, par. 2, lines 1–5).

[6] Some modern readers have had the same problem. For example, Westfall [1] writes "Newton recognized that Boyle's law implies the dynamic condition of simple harmonic motion" (p. 497). Similarly, Truesdell [2] writes that "In all three [propositions] the condensation is taken as proportional to the displacement, . . . It is plain that he has not yet hit upon the idea of relative displacement . . ." and that "Newton does not use any relation between the pressure and the density of the air." (p. XXII and p. XXIII).

# 3. Taylor (1713)

Brook Taylor presented his work on the vibrating string to the Royal Society in September of 1712. His paper appeared in 1714 in the volume of the *Philosophical Transactions* for 1713;[1] and a revised version was included in his book on the calculus,[2] published in 1715. This work introduces the *pendulum condition*. All elements of the string are supposed to undergo small vibrations as simple pendulums all of the same period; hence they are restored by *harmonic forces* of the same intensity. From a static analysis, the force on an element is found to be proportional to the curvature. The equality of this force with the harmonic force is the pendulum condition. From this, the shape of the string and the frequency of vibration are determined. In 3.1, we will give an exposition of this work; in section 3.2, we will quote Taylor's unpublished description of experiments on absolute frequency; and in section 3.3, we will make some other remarks on Taylor's background and influence.


## 3.1. Vibrating String

Let a string be stretched between two points of the $z$-axis in the $z$–$y$ plane. At a given instant of time it will lie along a curve $\mathbf{X}(u) = (z(u), y(u))$. Let $s(u)$ denote arc length along the curve. Differentiation with respect to $u$ and $s$ will be denoted respectively by a dot and a slash; $\dot{s}$ is to be constant (as a function of $u$) and a given value of $u$ is to specify a given point of the string no matter how it is stretched.[3] Let $P$ denote the tension of the string and let $\rho$ denote the mass per unit length. Note that $P$ and $\rho$ depend on how much the string has been stretched from its equilibrium length; but they can be considered as constants in the limit of

---

[1] Taylor [1].

[2] Taylor [2], lemmas VIII and IX, and propositions XXII and XXIII, pp. 88–93. (Second edition.)

[3] Of course, Taylor does not use vector notation. The parameter $u$ is suppressed by Taylor. He does not make explicit use of derivatives with respect to $s$ but they are implicit in his geometric arguments.

small vibrations. Taylor considers each element[4] $ds$ of the string to be a particle which should be treated very much as a particle of mass $\rho\,ds$ in the mechanics of one degree of freedom. The force acting on it is "by the principles of mechanics,"[5] $P[\mathbf{T}(s+ds)-\mathbf{T}(s)]=P\mathbf{T}'\,ds=P\kappa\mathbf{N}\,ds$, where $\mathbf{T}=\mathbf{X}'$ is the unit tangent vector, $\mathbf{N}$ is the unit normal obtained by rotating $\mathbf{T}$ ninety degrees counterclockwise, and $\kappa$ is the curvature.[6]

At this point, the modern reader may be tempted to presume that Taylor will use the momentum principle to obtain the equation of motion which in the small vibration limit becomes the wave equation.[7] But Taylor needs to begin with some supposition about the motion and only afterwards, effectively by the momentum law, can he obtain extra information.

The main thing that he presumably knows about the motion is Mersenne's law (1.1) with the notion of isochronism. He sees each element $\rho\,ds$ as a simple particle oscillating in one dimension with the same period $T$ no matter which the element and what the amplitude of the oscillation. Now, presumably thinking of this from the perspective of dynamics in a single degree of freedom, Taylor concludes that each element $\rho\,ds$ moves as a simple pendulum of length $\mathfrak{a}\mathfrak{g}$, in our notation, given by (1.4). To further simplify the situation he makes the assumption of simultaneous crossing of the axis (and in fact restricts himself to the case in which the string approaches the axis from one side only). Thus, at any instant, with the string displaced to the curve $(z(u),\ y(u))$, Taylor has that the element $\rho\,ds$ is restored by the harmonic force[8] $-\mathfrak{a}^{-1}\rho y\mathbf{N}\,ds$. This he equates to the force obtained previously to obtain the condition

$$\hat{\mathfrak{a}}\kappa=-y \quad \text{where } \hat{\mathfrak{a}}=\mathfrak{a}P/\rho. \tag{3.1}$$

---

[4] Taylor doesn't use differential notation. Here is a paraphrase of his description: At any instant of its vibration, let a tense cord, stretched between the points A and B, take any form of a curve ApP$\pi$B [for which a figure is given]. Conceive the string to consist of equal rigid particles, infinitely small, as pP and P$\pi$, etc. Taylor doesn't introduce a notation for the mass density $\rho$; geometrically Taylor's pP is $ds$ and physically it is $\rho\,ds$.

[5] (Changed to "principles of statics" in the *Methodus incrementorum*.) In Taylor's terminology: "[the] force, by which one particle pP is urged, will be to the tension of the string . . . as pP/PR . . . the acceleration is as $1/PR$, that is, as the curvature at P" (lemma 2). This force is obtained also by Jakob Bernoulli in his work on the elastica, 1691–1694 (Truesdell [1], pp. 88–96).

[6] Note on the calculus of curvature: $\mathbf{T}=\mathbf{X}'$ is a unit vector since $s$ is arc length and so $\mathbf{T}'$ is perpendicular to $\mathbf{T}$; writing $\mathbf{T}'=\kappa\mathbf{N}$ defines the curvature $\kappa$ and the normal $\mathbf{N}$ up to a sign convention. The center of curvature of $\mathbf{X}$ at $\mathbf{X}_0=\mathbf{X}(s_0)$ is $\mathbf{m}_0=\mathbf{X}_0+(1/\kappa)\mathbf{N}_0$ which is easily seen because the first two derivatives of $|\mathbf{X}-\mathbf{m}_0|^2$ at $s=s_0$ are zero; hence $1/\kappa$ is the radius of curvature at $s=s_0$. Now if $\mathbf{X}=(z,y)$, then $\mathbf{T}=(z',y')$, $\kappa\mathbf{N}=(z'',y'')$, $z'^2+y'^2=1$ and $z'z''+y'y''=0$; thus $\kappa^2=z''^2+y''^2=z''^2/y'^2$ and for an appropriate sign convention $-\kappa=z''/y'$ and by the chain rule, $-\kappa=\ddot{z}/\dot{s}\dot{y}$ because $\dot{s}$ is constant. Finally, if $\mathbf{T}=(\cos\alpha,\sin\alpha)$, $\mathbf{T}'=\alpha'(-\sin\alpha,\cos\alpha)$ and so $\kappa=\alpha'$.

[7] Indeed, Struik [1] writes that "This means that for small vibrations Taylor has in principle [the wave equation, although] there is no evidence that he had any notion of partial derivatives." (p. 352)

[8] This is an idealization of what Taylor actually presents. He first treats the difficulties that we come to in the next paragraph; and as a result of this finds that the force $P\kappa\,ds$ is

Aside from the fact that Taylor has not yet taken the small oscillation limit, this is the *pendulum condition*. It becomes central to the subject.

There would be difficulties with this equation if one wanted it to hold for vibrations that are not small: $y$ would not be the length of the arc along which the string element is actually displaced and the equation is not invariant in time. For, if $y(s, t)$ varies in time as an harmonic pendulum of period $T$ for all $s$, then $y(s, t) = \gamma(t)y(s)$, where $y(s)$ is the position at some given instant, and the equation (3.1) could hold only if $\kappa(s, t) = \gamma(t)\kappa(s)$. Taylor must have perceived these difficulties as his main problem because he begins his paper with a lemma that shows that they don't occur in the limit of small vibrations. Taylor gives a geometrical argument but it is essentially this: Write $\mathbf{T} = (\cos \alpha, \sin \alpha)$ so that $\kappa = \alpha'$; if $\beta$ is the corresponding angle for the curve $(z(s), \gamma y(s))$, then $\sin \beta = \gamma \sin \alpha$; in the limit of small vibrations, $\alpha$ and $\beta$ are small and so $\beta = \gamma\alpha$ and furthermore $s$ becomes the arc length along both curves; so $\beta' = \gamma\alpha'$ gives the needed relationship between the curvatures.

Now, things would be simplest if the small vibration limit of (3.1) were taken directly. However, this equation was of some importance in its own right, having been studied by Jakob Bernoulli in his work on the elastica. As Bernoulli had done, Taylor substitutes $\ddot{z}/\dot{s}\dot{y}$ for $-\kappa$ to put the equation in the form $\hat{a}\ddot{z} = \dot{s}y\dot{y}$ which he integrates to obtain

$$\hat{a}\dot{z} = \tfrac{1}{2}\dot{s}[y^2 - c^2 + 2\hat{a}],$$

where he writes the constant of integration so that $\dot{z} = \dot{s}$, and therefore $\dot{y} = 0$, when $|y|$ takes on its maximum value, $c$. Since $\dot{s}^2 = \dot{z}^2 + \dot{y}^2$, he eliminates it (as Bernoulli had done in other notation) to obtain the equation

$$\dot{z} = \pm \frac{[2\hat{a} - c^2]\dot{y} + y^2\dot{y}}{[4\hat{a}c^2 - 4\hat{a}y^2 - y^4 - c^4 + 2c^2y^2]^{1/2}}. \tag{3.2}$$

In his book he covers the applications of (3.1) or (3.2) to describe the shape of a cylindrical sail filled with water and, with a change in the sign of $\hat{a}$, to describe the shape of an arch that supports a fluid. In any case, it is only in the form (3.2) that he takes the small vibration limit, $|\dot{y}| \ll \hat{a}$ (which implies that $c = |y|_{max} \ll \hat{a}$). In the limit, (3.2) goes into

$$\dot{z} = \pm\sqrt{\hat{a}}\dot{y}[c^2 - y^2]^{-1/2} \tag{3.3}$$

proportional to displacement, though not necessarily with the same constant of proportionality for all elements $ds$. Then he says that the constants of proportionality must be the same for all elements $ds$, obtaining (3.1), which in his notation is the condition that "the radius of curvature at E [an arbitrary point on the string] will be ... $aa/E\eta$," where $E\eta$ is our $y$, and $aa$ is our $\hat{a}$.

whose integration is familiar geometry; and we can write, if we use functional notation for convenience, that

$$y = \pm c \, \sin \frac{z}{\sqrt{\hat{a}}} \, . \tag{3.4}$$

Taylor assumes that $|y|$ has a unique maximum at the mid-point and of course that $y = 0$ at the end points which we can take to be at $z = 0$ and $z = l$ of the $z$-axis. It follows that $\sqrt{\hat{a}} = l/\pi$ and hence that

$$\mathfrak{a} = \frac{l^2}{\pi^2} \frac{\rho}{P} \, . \tag{3.5}$$

That is, the *simple isochronous pendulum* has length[9] $\mathfrak{a}\mathfrak{g}$, with $\mathfrak{a}$ given by (3.5); the frequency and period are given by (1.3). Thus Taylor has derived Mersenne's law (1.1) with its constant[10]

$$\nu = \frac{1}{2l} \sqrt{P/\rho} \, . \tag{3.6}$$

But in the course of obtaining this famous result, Taylor encounters a paradox that will not be correctly solved until mid-century: He has that isochronism and simultaneous crossing of the axis imply the pendulum condition which in turn implies that the shape of the vibrating string at any instant is that of a sine curve; but it is also clear to Taylor that a string that is plucked at its center, for example, should have the properties of isochronism and simultaneous crossing of the axis though it is certainly not a sine curve at the time of its release.

Taylor's solution to this paradox is to imagine that after a small number of vibrations the string will adjust itself to the proper shape. He regards this adjusting process to be beyond the scope of mathematical description; yet he attempts a qualitative description of it. And in attempting this qualitative description he actually makes a vague use of the *momentum principle*, though this use mostly shows how unavailable it was to him: He argues, more or less (and his two versions differ), that the points of the string at which $\hat{a}|\kappa| > |y|$ are those points for which $|y|$ is too large and those at which $\hat{a}|\kappa| < |y|$ are those for which $|y|$ is too small for the string to have the shape of a sine function; so that the acceleration, which is proportional to $|\kappa|$, will tend to bring the string into the shape of a sine function. The argument will not often be referred to but the solution will be accepted until mid-century.

---

[9] Taylor refers to Newton's *Principia* for the period of a pendulum. In Taylor's notation the length is $Naa/PL$, where $L$ is our $l$, $N$ is our $\rho\mathfrak{g}l$, and $aa$ is our $\hat{a}$.

[10] In Taylor's terminology: "The number of [semi] vibrations of the string, in the time of one [semi] vibration of the pendulum $D$, is $(c/d)\sqrt{(P/N)\times(D/L)}$," where $c/d$ is $\pi$ and $D = \mathfrak{g}/\pi^2$ is the length of a seconds pendulum whose period is two seconds.

## 3.2. Absolute Frequency

In March $1712/13^{11}$, about six months after presenting his paper to the Royal Society, Taylor measured the frequencies of vibrating strings. Among his manuscripts at Cambridge[12] are the following notes:

6 March 1712/13

I applied a quill to the crown wheel [escape wheel] of my chamber clock, and making it fast to one of the [pillars] of the clock, I let the works run down for 7 minutes and by my Harpsicord I found the quill to sound Alamire[13] in alt:, and by the works of the clock the quill struck 766 teeth per second.[14] By which means the quill made $2 \times 766 = 1532$ vibrations per second. For in striking each tooth the quill goes once forwards and once backwards.[15] Then with some wire which weighed 1 grain per foot, I made another experiment, hanging 10 ounces or 4800 grains to stretch it, and found it at 12.3 inches long to sound Alamire, two eighths [octaves] below the former. So that this wire ought to strike the air 383 times per second.[16] But according to my second theorem [in] *De Motu Nervi* this wire made 383 vibrations per second, which agrees wonderfully with the experiment.

9 March 1712/13

I repeated the same experiment, and found that 2341 vibrations per second made Elami above the compass of my harpsicord. Which being a fifth above A, (the note in the former experiment,) that note should make $1561 (= 2/3 \times 2341)$ vibrations per second, which is about 1/52 part bigger than in that experiment: a difference scarce worth regarding.

10 March 1712/13

I stretcht a string of the best white packthread 126 feet long, weighing 350 grains, with a weight of 52 ounces or 24960 grains. This according to my second [theorem in] *De Motu Nervi Tensi* ought to make 4.27 vibrations per second. But by observing the vibrations of it for 5 seconds at a time—thrice repeated, I found it to make but 4.07 vibrations in a second. Which is about 1/20 less than it should

---

[11] That is, March of 1713 by the new style (Gregorian) calendar. In England the old style (Julian) calendar, with the new year beginning on March 25, was still in use.

[12] We would like to thank N. C. Buck of the St. John's College Library, Brook Taylor Collection (class mark U.19) for help in determining which manuscripts were relevant and for sending copies. We would also like to thank Prof. Philip S. Jones for information about the contents of Taylor's notes on music and vibration.

[13] An *A* in the notation of the hexachord system.

[14] Were the pendulum of Taylor's chamber clock seven inches long, the pallets would release a tooth of the escape wheel each .84th of a second or 4,320 teeth per hour; but with the pallets removed, the escape wheel turned at a rate of 766 teeth per second for a total of 321,720 teeth in seven minutes while the clock ran down. (From this we could conclude that the clock needed winding once each 74 hours.)

[15] Taylor's frequencies are double frequencies. A typical frequency for an *A* in early eighteenth century London seems to have been around 420 (Ellis [1], pp. 496–497), relative to which Taylor's harpsichord was between a semitone and a tone flat.

[16] Thus, $P = 4800$ grains, $\rho = \frac{1}{32}$ grains sec$^2$ ft$^{-1}$, and $l = 1.025$ ft. By (3.6), $\nu = 191.2$ cycles per second or 382.4 semi-cycles per second.

be by the calculation. This difference is much less than what it was in the experiments I tried with strings of $24\frac{1}{2}$ feet. Which however agreed very well with each other, making all of them between 72 and 73 vibrations in the time they ought to make 100; which is more than 1/4 part less than the calculation. Whence it appears that the more the string is stretcht the nearer it agrees with my calculation, as appears by the 2 experiments made with my clock (7 and 9 th March).

## 3.3. Remarks

We will follow Taylor's influence in the later sections on the Bernoullis and Euler. However, Taylor's work also gained a wider fame. For example, Maclaurin, in his book of 1742 on calculus and its applications, gives a brief presentation.[17] In 1748, Diderot published a detailed presentation in his early *Mémoires sur Differents Sujets de Mathématiques.*[18] Robert Smith, who succeeded his cousin Cotes as professor of astronomy at Cambridge, gave a detailed exposition of Taylor's argument in his *Harmonics* of 1749.[19] D'Alembert's article on the vibrating string for the *Encyclopédie* of 1766 was essentially centered on Taylor's work.[20] Daniel Bernoulli, in his mid-century debates with d'Alembert and Euler, leaned on Taylor for influence even to the point of attributing to Taylor his own analysis of higher modes.[21] Thus, Taylor's vibrating string was at the height of its fame, taking its place as part of the Enlightenment half a century after its publication. Even the 1803 edition of the *Encyclopedia Britannica* included a detailed presentation of Taylor's analysis.[22]

There can be no doubt that music was an inspiration for Taylor in his interest in the problem of the vibrating string. We know, at least, that he came from a musical family.[23] Although Taylor did not publish anything on music, there are more than a hundred pages of his notes on music theory in the Cambridge collection, among which there are some comments on physical foundations of music. He makes a number of remarks in the seventeenth century tradition of Music and Vibrations, e.g. on the quantification of pitch by frequency, on the nature of consonance, and on the lower limit of audible frequency. He also pursued music in the grand Pythagorean tradition. In his notes there are comments on the numerology

---

[17] Maclaurin [1], II, sections 926–927, pp. 743–747.

[18] Diderot [1], pp. 397–412.

[19] Smith [1], pp. 247–265. For certain initial conditions when the shape of the string is not sinusoidal, Smith applies Taylor's argument for the solution of the paradox but only to show that the string still has the properties of *isochronism* and *simultaneous crossing of the axis.*

[20] D'Alembert [1].

[21] Daniel Bernoulli [11], pp. 147–149.

[22] Robison [1].

[23] Young [1], Taylor's grandson, writes that Taylor's father often entertained leading musicians of the day. See also, Jeans [1].

of musical intervals; he writes that "Music is one of those arts ... which have their foundation in certain unvariable principles taken from the nature of things." It seems that Taylor saw his own work as a culminating point in this tradition; in his manuscript essay "Of Music" (see below) Taylor writes: "the ancients supposed it, and it has now been demonstrated, that if a string of a uniform thickness be stretched by a given force, the swiftness with which any part [between bridges] vibrates is reciprocally proportional to the length of that part." (In the revised presentation of his analysis of the vibrating string that he published in his book, Taylor refers to the string as a "musical string"; in the third corollary he remarks that the law, $\nu \propto 1/l$, is the basis for the Pythagorean ratios.)

It is surprising that Taylor does not even mention higher modes. Their existence was certainly known in Taylor's time; and Taylor should have been acquainted with them through practical experience with music. Furthermore, Francis Robartes, who had published a paper on the connection between harmonics and nodes,[24] was an active member of the Royal Society in Taylor's time. In principle, Taylor could have regarded his analysis as applicable to a pure higher mode; but it does not seem that he did so. In fact, looking at the shape curve (3.4) from a purely *geometric* point of view, one does not see the possibilities of a periodic function! Taylor's solution of his paradox rules out the possibility of superposition.

Finally, we would like to point out some circumstantial evidence that Newton took an interest in Taylor's analysis of the vibrating string: Taylor had contact with Newton at least from 1712 at the Royal Society,[25] where Newton was president. Taylor himself was secretary from 1714 to 1718. Newton included Taylor on his list of about ten people mentioned by name to whom copies of the 1713 edition of the *Principia* were to be sent.[26] It seems quite possible that Taylor and Newton also had a more casual friendly relationship.[27] Among Taylor's manuscripts on music there is a title page "On Musick by Dr. Brook Taylor, Sir Isaac Newton and Dr. Pepusch[28]" and there are three manuscripts, one "Of Music" clearly by Taylor, another "Of Music from Dr Pepusch," and a third without title which we presume to be written by Newton. The style seems right; it is concise with numbered

---

[24] Robartes [1].

[25] In 1722 Taylor was appointed to the Society's committee investigating the charges of Leibniz against Newton. (Newton [2], p. xxv.) In the same year he performed (with Hawksbee) experiments on capillarity at a meeting of the Society, with Newton in the chair (*ibid.* p. 396.)

[26] Cohen [1], p. 247. Most of the seventy copies listed were to be sent to institutions, including libraries and royalty.

[27] William Stukeley wrote that in 1717–1718 he often visited Newton, sometimes with Brook Taylor and others. (Cohen [1], p. 300.) See also Manuel [1], p. 252.

[28] This is, no doubt, John Christopher Pepusch (1667–1752), popularly known as the arranger of the songs for *The Beggar's Opera*. He received an honorary degree at Oxford in 1713. Pepusch was known for his esoteric knowledge of ancient music theory. He was elected a fellow of the Royal Society for a paper on ancient genera in 1745 (*Grove's Dictionary of Music & Musicians*, fifth edn.)

paragraphs and capitalized nouns. For example, paragraph two reads: "A Musical Sound or Note is when the Vibration is regular and uniform, the several strokes or pulses of the Air also following each other too swift to be distinguished by sense," which can be contrasted with Taylor's corresponding statement: "... it must be observed that sound is produced by a vibrating or trembling motion of bodies which makes an undulation in the air, which affecting the organ of hearing is the cause of sound." Newton had a strong interest in the Pythagorean tradition and, at least at one point, he believed not only that the ancients knew Mersenne's law but that they knew its application to planetary motion and thence, implicitly, the inverse square law of gravitation.[29]

---

[29] See McGuire & Rattansi [1] and Dostrovsky [1], p. 211.

# 4. Sauveur (1713)

Joseph Sauveur presented his derivation of the vibrational frequency of a musical string to the Paris *Académie Royale des Sciences* in 1713. The paper was published[1] shortly after his death in 1716. In his derivation, Sauveur supposes that the string is stretched horizontally in a gravitational field and that it undergoes small horizontal vibrations. With the vibrations assumed small even compared with the sag in the string, Sauveur obtains the situation in which the string vibrates in a swinging motion. Somewhat implicitly he makes the simplifying assumption that the string undergoes this swinging motion as a rigid body. Thus he succeeds in reducing the problem to one that is already familiar. He emphasizes that in general the gravitational field has a negligible effect on a taut vibrating string and thus that his result for frequency holds in general.

It seems hard to avoid the first impression that the negligible effect of the gravitational field guarantees also the irrelevance of Sauveur's argument. But the argument is correct except for the assumption of rigidity which restricts attention to the fundamental mode but introduces an error of less than $\frac{2}{3}$%. Of course, in reducing the problem to a familiar one in a single degree of freedom, Sauveur does not advance the understanding of dynamics. We will begin by presenting his argument in a form that is slightly idealized.[2] This we will qualify by indicating some of his techniques of calculus. In section 4.2, we will make a few remarks on the experimental confirmation that he already had and on the reception of his work.

## 4.1. Vibrating String

Let the string, having a uniform linear density $\rho$, be stretched in $z$–$x$–$y$ cartesian space along the $z$-axis from $-l/2$ to $l/2$ with a tension $P$ at the end points. Let the gravitational field $g$ act in the direction of the negative $x$-axis. In the limit of $g$ small, or as Sauveur prefers $g\rho l/P$ small, it can be

---

[1] Sauveur [3, 4].

[2] In section 4.1 we expand upon the presentation given in Dostrovsky [1], section 3.5.

assumed that the string hangs in a parabolic curve

$$x = f\left[ -1 + \frac{4}{l^2} z^2 \right] \tag{4.1}$$

(and $y = 0$) in equilibrium, where $f$ is the small sag in the string which Sauveur calls the *flèche*. The other constant is chosen so that $x = 0$ when $z = \pm l/2$.

To determine $f$ in terms of $P$, $\rho$, $l$, and $\mathfrak{g}$, Sauveur simply uses the fact that the vertical components of the forces of tension at the end points are balanced by the weight of the string. In the limit of small $f$, arc length along the string is given by $z$; so the vertical forces exerted at the end points are $\pm P \, dx/dz$, evaluated at the end points, respectively. The weight is $\mathfrak{g}\rho l$. Thus by (4.1),

$$f = \frac{\mathfrak{g}\rho l^2}{8P}. \tag{4.2}$$

Now, Sauveur considers oscillations of the string in the $y$-direction that are small compared with $f$. Thus the string undergoes a swinging motion in the gravitational field.[3] Implicitly, Sauveur assumes that it will undergo this swinging motion as a rigid body; thus he has explicitly that the *length of the isochronous simple pendulum*, $\mathfrak{ag}$, is the distance from the $z$-axis to the center of oscillation of the uniform curve (4.1)

$$\mathfrak{ag} = -\frac{\displaystyle\int_0^{l/2} x^2 \, dz}{\displaystyle\int_0^{l/2} x \, dz} = \tfrac{4}{5} f \tag{4.3}$$

in the limit of small $f$ since $z$ is being used in the place of arc length.

From (4.2) and (4.3) we have in our standard notation that

$$\mathfrak{a} = \frac{1}{10} \frac{\rho l^2}{P}. \tag{4.4}$$

The gravitational field $\mathfrak{g}$ has cancelled out of the problem! (This would be the correct result for the ideal string's fundamental frequency if 10 were

---

[3] It is easily checked that the theory of the swinging taut string agrees with the theory of the ideal vibrating string. We can show this simply by referring to the theory of the hanging chain of Daniel Bernoulli and Euler (section 9.7). As a function of $\hat{x} = x + f$, we have that the string has a density proportional to $1/\sqrt{\hat{x}}$ since $z$ is arc length in the limit of small $f$. The *pendulum condition* is then given by (9.31) with $\nu = -1/2$ which has the solutions $y = \cos\sqrt{2\hat{x}/\mathfrak{ag}}$, for the odd modes and $y = \sin\sqrt{2\hat{x}/\mathfrak{ag}}$, for the even modes, with the condition $\mathfrak{ag} = 8f/\pi^2 n^2$, $n = 1, 2, 3, \ldots$. From (4.2) and (3.5) one sees that this agrees with the standard theory of the ideal string. The odd modes are given in (9.37) but the even modes correspond to a chain that is fixed at the bottom. (The hanging chain is not a model that describes both even and odd modes together.)

replaced by $\pi^2$.) The corresponding frequency[4] is

$$\nu = \frac{\sqrt{10}}{\pi} \frac{1}{2l} \sqrt{P/\rho} \qquad (4.5)$$

which is incorrect only by the factor $\sqrt{10}/\pi = 1.00658 \dots$.

Sauveur's argument for (4.2) differs in that he doesn't start with the limiting curve (4.1) but, by the balancing, determines that

$$\left.\frac{dx}{dz}\right]_{z=l/2} = \frac{g\rho l}{2P} \qquad (4.6)$$

whence that the tangent line to the end of the curve intersects the $x$-axis at[5] $-g\rho l^2/4P$. Then by a geometrical argument that is at least intuitively clear he argues that this is twice $-f$ in the limit in question. Sauveur's determination of $\alpha g$ in (4.3) differs in spite of the fact that he suggests that he could work with a parabolic curve. Instead he works with an arc of a circle. Nothing was more familiar than a circle in the calculus of the day! Let the arc be $u = \cos\alpha$, $v = \sin\alpha$, for $-\alpha_0 \leq \alpha \leq \alpha_0$, in the $u$–$v$ plane. Write $u_0 = \cos\alpha_0$, $v_0 = \sin\alpha_0$. Then, up to scaling $-x = u - u_0$, $z = v$, $l/2 = v_0$, and $f = 1 - u_0$. In the place of (4.3), Sauveur has[6]

$$p = \frac{\displaystyle\int_0^{\alpha_0} (u - u_0)^2 \, d\alpha}{\displaystyle\int_0^{\alpha_0} (u - u_0) \, d\alpha} = \frac{-3u_0v_0 + \alpha_0 + 2u_0^2\alpha_0}{2v_0 - 2u_0\alpha_0} \qquad (4.7)$$

which he sees easily from the geometry. He then seeks the limit of $p/f$ as $\alpha_0 \rightarrow 0$. First he evaluates this for a number of angles $\alpha_0$ and finds that even for the semi-circle the ratio is not far from $\frac{4}{5}$. But he has to work fairly hard to obtain the limit. His first step is essentially to use l'Hospital's rule.

---

[4] In Sauveur's notation the double frequency is $\sqrt{10}\, apq/\sqrt{cn^2}$, where $n$ is our $l$, $c/a$ is $\rho g$, $p$ is $P$, and $q = g/\pi^2$, the length of a second's pendulum whose period is two seconds.

[5] In Sauveur's terminology: "By the rules of mechanics the weight $cn/a$ of the string ADB ($n$) is to the weight $P$ ($p$) that stretches this string as the small diagonal EF ($2x$) is to the tangent EB ($t$), therefore $cnt = 2apx$," where $x$ is our $f$, $n$ is our $l$, $c/a$ is our $\rho g$, and E is the intersection of the tangent at the end point with the bisecting vertical (par. 14).

[6] In Sauveur's notation:

$$p = \frac{\displaystyle\int y^2 \, du}{\displaystyle\int y \, du} = \frac{ar^2 + 2ab^2 - 3bcr}{2cr - 2ab},$$

where $y$ is our $u - u_0$, $du$ is $d\alpha$, $a$ is $\alpha_0$, $b$ is $u_0$, $c$ is $v_0$, and $r$ is the radius of the circle which we take as unity, (pars. 39–41).

This with substitutions[7] $du_0/d\alpha_0 = -v_0$, $dv_0/d\alpha_0 = u_0$, $u_0^2 = 1 - v_0^2$, and $u_0 = 1 - f$ gives

$$\lim_{\alpha_0 \to 0} \frac{p}{f} = \lim_{\alpha_0 \to 0} \frac{2v_0 - 2\alpha_0 + 2f\alpha_0}{v_0 - \alpha_0 + 2f\alpha_0}. \tag{4.8}$$

But next he effectively applies l'Hospital's rule to the expression $(\alpha_0 - u_0 v_0)/2fv_0$ (a ratio that is motivated geometrically by his interest in centers of mass) to obtain, with the help of similar substitutions, the approximation

$$\alpha_0 \approx \frac{3v_0 - fv_0}{3 - 2f} \tag{4.9}$$

which he substitutes into (4.8) to see the desired result, $\frac{4}{5}$.

## 4.2. Remarks

Sauveur's foremost interest was in understanding the nature of musical sound.[8] He knew Mersenne's law, which in terms of the *length of the simple isochronous pendulum* is the proportionality

$$\mathfrak{ag} \propto \frac{\mathfrak{g}\rho l^2}{P}. \tag{4.10}$$

(See (1.1), (1.3), and (1.4).) Also, he was concerned with determining absolute frequency or equivalently with finding the constant of proportionality in (4.10). In 1700 he presented the Paris *Académie* with a method for determining absolute frequency experimentally.[9] In this method a pair of organ pipes is tuned accurately to a close interval and the beats produced in a given period of time are counted. Thus, Sauveur was in effect in possession of the equality (4.4) already in 1700, indeed with the constant $\frac{1}{10}$, if his experimental result were approximated accurately enough by a simple fraction. Now, the catenary problem was current in Sauveur's time and he could well have made some notes on the subject.[10] With his interest in musical sound, he would have been concerned with the monochord and, in particular, the taut catenary; thus he could have discovered the relation (4.2) for the *flèche* quite by chance. Then on the basis of (4.10) he would have seen immediately the proportionality $\mathfrak{ag} \propto f$ and on the basis of his experimental result he would have known the constant $\frac{4}{5}$ given in (4.3).

---

[7] In Sauveur's notation: $\text{BL}(da) \cdot \text{BN}(r) :: \text{BM}(db) \cdot \text{BC}(c) :: \text{LM}(dc) \cdot \text{NC}(b)$. That is, $da/r = db/c = dc/b$.

[8] Sauveur and his studies of music and sound are discussed in Dostrovsky [1], sections 5.3 and 6.2. and in Dostrovsky and Cannon [1].

[9] Sauveur [1, 2].

[10] That Sauveur took an interest in current problems is clear. For example, in 1696, he attempted a solution of the then-current brachistochrone problem, a solution which l'Hospital sent to Johann Bernoulli (Spiess [1], pp. 333–342.)

Thus Sauveur could have been in the position of trying to explain the constant $\frac{4}{5}$ before making any considerations of the physics or mechanics of the relevance of the proportionality. Finally, the problem of the center of oscillation was also current in Sauveur's time (for example, Johann Bernoulli had written the Paris *Académie* a letter[11] containing the formula that Sauveur used. Thus, the means for obtaining the constant $\frac{4}{5}$ were very much at hand. Because Sauveur does not explain the physics of the vibration that he considers, it is conceivable that he obtained his result by mere association and not by physical reasoning. It seems that some of his readers presumed this to be the case:

For example, in a very out of date article by John Robison, in the Supplement to the 1803 edition of the *Encyclopedia Britannica*, supposedly on the Trumpet Marine, there appear after a praising presentation of Taylor's argument the remarks:[12]

... The merit of Dr. Taylor is not sufficiently attended to ... such as have occasion to speak of the absolute number of vibrations made by any musical note, always quote Mr. Sauveur.... [Besides his experimental determination, he has] given a mechanical investigation of the problem, which gives the same number of vibrations that he observed ... this demonstration ... is a mere paralogism where errors compensate errors; and the assumption on which he proceeds is quite gratuitous, and has nothing to do with the subject.

Of more relevance for our present study is Daniel Bernoulli's letter to Euler of December 24, 1726 in which he approves Sauveur's experimental determination in principle only and asserts, without reasons, that he "altogether disapproves" of Sauveur's derivation of the vibrational frequency of a musical string.[13] On the other hand, Cramer in his *De Sono* of 1722 (Chapter 6), accepts Sauveur's derivation, probably without understanding it, saying that "he displayes no little skill in the more recondite geometry."

Sauveur's experimental determination was not overlooked. Also, he was quite famous throughout the eighteenth century for his demonstration and qualitative explanation of the existence of the higher modes of a vibrating string. Euler refers to him for this in his *Tentamen Novae Theoriae Musicae*[14] of 1739 as does Lagrange[15] in 1759 and also Diderot[16] in 1748, and especially Rameau, whose ideas on music theory, beginning in 1726, were influenced by the knowledge of harmonics.[17]

---

[11] Johann I Bernoulli [2].

[12] See footnote 22, Chapter 3.

[13] Bernoulli's comments are partially quoted by Truesdell [1], p. 143. We would like to thank Prof. Walter Habicht and Beatrice Bosshart of the University of Basel for providing a typescript of the letter.

[14] Euler [7], chap. 1, par. 41.

[15] Lagrange [1], p. 45.

[16] Diderot [1], pp. 445–456.

[17] Cf. Rameau [1], p. 17.

# 5. Hermann (1716)

Jakob Hermann published his analysis of the propagating pressure wave in his *Phoronomia*[1] of 1716. In the same year he published his analysis of the vibrating string in the *Acta Eruditorum*.[2] In both of these works he seeks and finds simple harmonic motions associated with the respective problems; in both cases his analysis is wrong. This demonstrates quite well how great confusion about dynamics in many degrees of freedom was compatible with a sophisticated understanding of dynamics in a single degree of freedom. We will present these works in sections 5.1 and 5.2 respectively, following which we'll make some general remarks and discuss Hermann's treatment of simple harmonic motion in section 5.3.

## 5.1. Pressure Wave

Hermann's pressure wave propagates through a one dimensional continuum taken somewhat loosely to be air along the real line. The wave is initiated by a vibrating body which at time $t = 0$ causes the air in the interval $(-\varepsilon, \varepsilon)$ to be compressed to a density $\rho_0$ while the air outside is at its equilibrium density. Hermann needs to know the wave length $\lambda$ with which the propagation is to occur and he assumes that the air at the points $n\lambda/2$, $n = 0, 1, 2, \ldots$, will not move[3] (except initially when it may move a distance $\varepsilon$). Let $x(t)$ label the position of the air that is at $\varepsilon$ at time[4] $t = 0$; this air is to be thought of as a particle "$\mu$" whose mass $\mu$ is left ambiguous. Before it is possible to say what the forces on $\mu$ are, it is necessary to give a general description of its motion: Being between regions of different densities, $\mu$ experiences a force to the right; so it is accelerated to the right compressing the air that was originally in $(\varepsilon, \lambda/2)$ in front of it while the

---

[1] Hermann [1], chap. XXIII, pp. 373–376.

[2] Hermann [2].

[3] Johann II Bernoulli and even Daniel Bernoulli will also have nodal points in their descriptions of a propagating pressure wave. (See Chapter 11.)

[4] In the figure, redrawn from the *Phoronomia*, AB = CD, E and F are the midpoints of AB and CD, 0 is the midpoint of EF and the center of the circle ENF. EM represents the density

air that was originally in $(0, \varepsilon)$ expands behind it; thus the pressure in front increases while the pressure behind decreases, balancing when $x = \lambda/4$ and by symmetry stopping $\mu$ at $x = \lambda/2 - \varepsilon$ when the air that was in $(\varepsilon, \lambda/2)$ is compressed to[5] $(\lambda/2 - \varepsilon, \lambda/2)$ while the air that was in $(0, \varepsilon)$ is expanded to $(0, \lambda/2 - \varepsilon)$; whence the process reverses itself except that the air in $(\lambda/2, \lambda/2 + \varepsilon)$, which is also compressed to the density $\rho_0$, expands to the right, propagating the wave. Now, to find the force on $\mu$, assume Boyle's law, $P = c\rho$, and suppose the air in $(0, x)$ and $(x, \lambda/2)$ always to have uniform densities, whence respective pressures $c\varepsilon\rho_0/x$ and $c\varepsilon\rho_0/[\lambda/2 - x]$; thus the force on $\mu$ is given by the difference[6]

$$2c\varepsilon\rho_0[\lambda/4 - x]/x[\lambda/2 - x]. \tag{5.1}$$

The momentum principle in one degree of freedom then gives Hermann the equality of this force with $\mu v \, dv/dx$ where $v = \dot{x}$ which Hermann integrates to obtain[7]

$$v^2 = \frac{c}{\mu} 2\varepsilon\rho_0 \ln x[\lambda/2 - x] + \text{const.} \tag{5.2}$$

with the constant chosen so that $v = 0$ at time $t = 0$. Actually he only asserts proportionality and he could do no more since the mass $\mu$ is not defined.

Hermann explains the fact that Newton found the elements of air to undergo simple harmonic motion as follows: In a very rough approximation,

---

of the air in **AB**. G is an instantaneous position of a particle moving to the right, and G0 = 0g. So in our notation, B is at $\varepsilon$, G is at $x(t)$, 0 is at $\lambda/4$, and F is at $\lambda/2$; EM represents $\rho_0$.

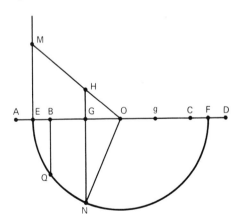

[5] To get the propagation started, Hermann has the air originally in $(\varepsilon, \lambda/2)$ compressed into $(\lambda/2 - \varepsilon, \lambda/2 + \varepsilon)$ since there is no other way to make this a high density region but he needs the symmetrical situation in which it is compressed to $(\lambda/2 - \varepsilon, \lambda/2)$; this adjustment is some sort of transient effect.

[6] In Hermann's notation: the force on a particle at G is proportional to **EM · EB**(1/**EG** − 1/**FG**).

[7] In Hermann's notation: $v^2$ is proportional to **EM · log** (**GN**$^2$/**BQ**$^2$).

he replaces $x[\lambda/2 - x]$ in (5.1) by $\varepsilon\lambda/2$; whence the force (5.1) becomes linear in $(\lambda/4 - x)$ and the momentum principle (5.2) becomes[8]

$$v \, dv = 4 \frac{c}{\mu} \frac{\rho_0}{\lambda} [\lambda/4 - x] \, dx. \tag{5.3}$$

Since, as we will mention in section 5.3, he has made a detailed study of simple harmonic motion earlier in the *Phoronomia*, Hermann now has immediately that[9]

$$v^2 = \frac{c}{\mu} 4 \frac{\rho_0}{\lambda} x[\lambda/2 - x] \tag{5.4}$$

and that the period is

$$T = \pi\sqrt{\mu/c} \sqrt{\lambda/\rho_0} \tag{5.5}$$

provided that he takes the amplitude of oscillation to be $\lambda/4$, rather than $\lambda/4 - \varepsilon$ (see (1.5)). Again, he can assert only proportionality since $\mu$ is not defined. Finally, it is clear from the description of his model that the velocity of propagation is

$$V = \lambda/T = \frac{1}{\pi} \sqrt{c/\mu} \sqrt{\rho_0\lambda} \tag{5.6}$$

or, as he says, "the velocities of the pulses will be as the square root of the interval of the pulses and the square root of the densities."

Hermann's propagating wave would look a little like a sloshing standing wave whose agitation propagates. Had Hermann associated pitch and loudness as well as velocity with his wave, the situation would have been as follows: The association of pitch with $1/T$ was well known. Wave length, though understood by Newton and Huygens, for example, was not generally understood; while loudness was generally associated with the degree of agitation. For Hermann this would mean that increasing loudness at a fixed pitch would mean increasing both $\rho_0$ and $\lambda$ and thus that the wave would propagate faster for louder sounds.

## 5.2. Vibrating String

Hermann analyses the vibrating string as follows: he assumes that the tension $P$ of a fixed section of a string varies linearly with its length[10] $L$,

---

[8] Hermann writes that the force on G is proportional to GH.

[9] In Hermann's notation: $v$ is proportional to $(GN \cdot EP)/EO$, where EP is the mean proportional between EM and EO $(EP = \sqrt{EM \cdot EO})$.

[10] In Hermann's notation: in the right triangle DFN the distance measured from D along DF represents the amount by which the displaced string is longer than the string in its equilibrium position and "the [corresponding] ordinates of the triangle DFN represent the excess of the tension of the string over the tension in [the equilibrium] position."

$\Delta P = c\Delta L$. Let $P_0$ be the tension of a stretched string of length $L_0$ in equilibrium; Hermann characterizes the displacement of the string by the single parameter $[L - L_0]$ and his main assumption is that it is restored towards equilibrium by the force $P - P_0 = c[L - L_0]$. With $M$ the mass of the string, he simply substitutes these quantities into the momentum principle for one degree of freedom: $M\,dv = c[L - L_0]\,dt$. Being familiar with simple harmonic motion (see (1.2) and (1.3)), Hermann sees at once that the period of vibration is

$$T = 2\pi\sqrt{M/c}. \tag{5.7}$$

Actually, Hermann doesn't work with an abstract constant $c$ but determines it right away by a careless extrapolation as $c = P_0/L_0$. Thus, he obtains Mersenne's law with a constant:

$$\nu = \frac{1}{2\pi}\sqrt{P_0/L_0 M}. \tag{5.8}$$

Though he presents his analysis as a response to Taylor's work, he does not note that his result differs by a factor[11] of $\pi$.

## 5.3. Remarks

Though wrong, both analyses do illustrate Hermann's familiarity with abstract simple harmonic motion. Newton and Taylor, for example, were thoroughly familiar with the cycloidal pendulum and recognized analogous dynamical situations; but Hermann may have been the first to abstract the dynamics to its equation[12] $a\ddot{x} = -x$ or rather $av\,dv = -x\,dx$ with $v = \dot{x}$ and he may have been the first to integrate this abstractly,[13] obtaining $av^2 = c^2 - x^2$; this in turn he studied systematically by standard geometrical methods.[14] The simplicity of his abstraction, however, is marred by his

---

[11] Hermann seems to have been ignorant of the fact that many reasonable estimates of the appropriate factor were available.

[12] In the *Phoronomia* (Book I, Section II, chap. 1, sec. 148) Hermann considers the motion that is produced when "gravitational forces are proportional to the distance from the center of gravity." He has already written Newton's law for "central or gravitational" forces both in the form $F = M(dv/dt)$ (sec. 131) and in the form $Fdx = Mvdv$ (sec. 132), with which he works.

[13] In the *Phoronomia* (*ibid.*) he obtains the velocity by geometrical methods and expresses the result as follows: "[When the scale of gravitational forces] is the straight line QCD drawn through the center D and making any angle QDA with the axis AD [that is, if E is the point on AD, such that CE is perpendicular to AD, then CE is the force at E], the scale of velocities will be the arc ASR of the ellipse whose transverse semi axis is AD and whose conjugate axis is DR, equal to the mean proportional between DA and QA." In his analysis of the vibrating string he rederives the result, as he says, for the convenience of the reader; but this time in *functional* notation: he integrates $-px\,dx = mludu$ obtaining directly $p(a^2 - x^2) = mlu^2$.

[14] In the *Phoronomia* (*ibid.*, sec. 149) and in the analysis of the vibrating string.

terminology, especially his reference to generalized "gravitational" forces, (as well as by his inappropriate applications).

Hermann refers to Lana's *Magisterii Naturae et Artis* of 1686 for an exhaustive treatment of the subject of taut strings,[15] though he says its author did not know sufficient geometry to deal with the problem of deriving Mersenne's law with its constant. Lana gives about sixty propositions on the subject; presumably they are indicative of Hermann's background in these matters. Some of the propositions make sense, including some that cover Mersenne's law and Hooke's law. Others are quite bizarre or wrong; for example, *Prop.* XXIV reads

The motions [velocities] of the points of a tense string, when it restores itself, are as their distances to the [nearest] immobile extremity.

Hermann presents his analyses as responses to the works of Newton and Taylor. He writes that "[Newton] dared to treat the theory of sounds geometrically, [although it] is surrounded with apparently insurmountable difficulties. [He] exhibited for us the most elegant theorem concerning the [simple harmonic] accelerations of pulses in elastic air." Hermann introduces his analysis as a simple version of Newton's: "[His] discussion, unless I am mistaken, reduces to this. . . ." Hermann finds Taylor's analysis to be deficient: "[Taylor] requires the shape of the vibrating string to be essentially restricted and he makes other dubious suppositions. For this reason I shall examine whether the period of an agitated string can be found without prior restrictions on the shape of its curve. . . ."

In spite of the fact that Hermann presents his analyses as responses to the works of Newton and Taylor, it is clear that he did not understand these authors. In particular, he did not understand the use of the momentum law in the case of many degrees of freedom. In the case of wave propagation this is clear enough; but in the case of the vibrating string we see him willing to apply the momentum principle for one degree of freedom as soon as he finds a mass, a force and something with the dimensions of distance, without regard for physical meaning! (Hermann's professional credentials were impeccable:[16] For example, he was a student of Jacob Bernoulli, a protégé of Leibniz, and throughout his life an active member of the Basel school. He preceded Euler (his younger relative) in a position at St. Petersburg.)

---

[15] Lana Terzi [1], Book VII, chap. 2, pp. 295–328.
[16] Cf. Fellman [1].

# 6. Cramer (1722)

Gabriel Cramer defended his thesis on sound[1] before the Geneva Academy in 1722. The thesis includes selected opinions of many authors,[2] from both the seventeenth century and antiquity. It attempts a presentation of Newton's analysis of the propagating pressure wave and it attempts a derivation of Mersenne's law. These attempts are by no means impressive; mainly, they are interesting in as much as they indicate the general understandings and confusions that would have been typical in the subject. We will discuss the thesis in section 6.1 and mention Cramer's influence in section 6.2.

## 6.1. Sound

Cramer attempts to demonstrate that sound consists in propagating pressure waves with "the motion of any particle of air ... analogous to the motion of the oscillating pendulum." He writes that Newton stands out among "the bolder or more expert philosophers [who] have ventured to define the motion of air [in the propagation of sound]," and he presents his own understanding of some of the ideas in Newton's analysis. He follows Newton in attributing simple harmonic motion of the air to the supposed simple harmonic motion of the source. Thus he avoids the confusion or the misunderstanding of many who believed or understood Newton to believe that the simple harmonic motion resulted from properties of the air itself. He writes that the vibrations of a string are the same as the oscillations of a pendulum and that the particles of air "imitate the motion of the tremulous body." Cramer refers to this description of sound as an hypothesis.

Cramer regards the rough agreement of Newton's calculated velocity of sound with experiment as a confirmation of the hypothesis. He reports Newton's velocity and refers the reader to the *Principia* for a mathematical

---

[1] Cramer [1].

[2] Cramer refers to Johann Bernoulli, Boyle, Carré, de la Hire, Fabri, Gassendi, Huygens, Lana, Kircher, Mersenne, Newton, Rohault, Sauveur, and others.

proof, "because it cannot be demonstrated physically [i.e. qualitatively]."
He mentions, without comment, Newton's suggestion that solid particles
in the air make the velocity higher than predicted. In avoiding consideration
of Newton's calculation, Cramer is satisfied with the significance of
Newton's preliminary result that the velocity is independent of amplitude
and frequency as an "important argument for the truth of our hypothesis."
To demonstrate this independence, Cramer presents three "cases," the
first two of which are apparently modelled after the first two cases in
the demonstration of Newton's *Prop.* XLVII (section 2.1). Cramer
states the same conditions and conclusions as does Newton; but he attempts
to make the reasoning more easily grasped by describing the motions and
forces as those of appropriate pendula. In his first case, Cramer follows
Newton without the precision—and without allowing the possibility of
misunderstood proportionalities:

### Case 1

We assume that two waves are of the same length but the particles in one cover
a longer distance in going and returning than in the other. Then the compression
in the first will be bigger, and so the force of elasticity will be larger. But also the
distance to be traversed will be larger. Therefore they compensate for each other.
As with the pendulum or the oscillating string, if the vibrations are to be larger,
the action of gravity or elasticity will be larger, smaller if they are to be smaller,
whence it is that the oscillations are isochronous. It is the same with waves of air;
the vibrations of any particle, whether they are larger or smaller, will be isochronous,
provided that the length of the wave remains the same. And since, in the time in
which the particles go and return the wave traverses its wavelength, the waves
traverse the wavelengths, which are equal, in the same time. Therefore the velocities
of both will be the same.

Cramer's second case comes out less fortunately; and his third case simply
draws the conclusion of independence from the first two.
    In his presentation of the vibrating string, Cramer is first concerned with
demonstrating that the motion is simple harmonic. He writes that

[the string] returns to the straight line by the action of elasticity, and its motion in
going is accelerated, and the action of the elasticity is as the string's distance from
the place of rest. For the more the string is displaced from that place, the more it
is curved and the greater is the action of the elasticity. The less it is curved, the
less is that action. Whence it follows that ... the vibrations of the string are
isochronous.

In purporting to derive part of Mersenne's law, namely $\nu \propto 1/l$, Cramer
writes with the freedom of a speculating philosopher:

The velocities of oscillating strings are as the velocities of pendulums; and these
are inversely as the squares of the periods, hence also the velocities of strings are
inversely as the squares of the periods. But these are also as the quantities of

motion divided by the quantities of matter. And the motions are as the restoring forces, which are as the curvatures of the string, but the curvatures are inversely as the lengths. The quantities of matter, when the strings are assumed to have the same thicknesses, are as the lengths. Therefore, dividing the inverse length by the length, it is found that the velocities must be inversely as the squares of the lengths. [The velocities], as we saw, [vary] according to the same proportion also with the periods. Therefore the periods are as the lengths. But the depths of tones are as the periods, hence they are as the lengths.

Like Hermann, Cramer does not know how to define forces and velocities in this situation involving many degrees of freedom. Thus, for example, he speaks of "the velocity of the string," meaning, at best, the maximum velocity of its mid-point, but he multiplies it by the mass of the entire string to obtain the "quantity of motion." Cramer's derivation of the rest of Mersenne's law is confused in a similar way.

Cramer discusses numerous other properties of sound and vibration, among which we find the following: He remarks on the problem of determining the constant in Mersenne's law, describing the two available experimental methods, Mersenne's and Sauveur's, and referring, without discussion, to Sauveur's theoretical derivation, but not mentioning Taylor. He describes the excitation by resonance of harmonics of a string and Sauveur's demonstration of the nodal points. He also discusses varieties of musical tones, musical intervals, consonance and dissonance, and the ear. He makes a few remarks on the production of sound by musical instruments, including pipes and bells. And he describes various phenomena associated with the propagation of sound in different media under different conditions.

## 6.2. Remarks

By itself Cramer's thesis probably would not have attracted much attention. However, Cramer himself was an established and well travelled mathematician of the period.[3] Though he never made any deeper contributions to vibration theory, Cramer did maintain his interest, as is evidenced by his correspondence in 1740 with Mairan, Secretary of the Paris *Académie*, which was published in 1741.[4] Mairan had proposed in 1720 and, in more detail in 1737, that air consists of particles of different elasticities which serve to propagate sounds of different frequencies.[5] Cramer explains to Mairan that in Newton's system "each particle of air makes exactly as many vibrations as the sonorous string," and that "in this

---

[3] Wolf [1], pp. 203–226.
[4] Cramer [1, 2].
[5] Mairan [1, 2].

system it is easy enough to understand how the same particle is capable of undergoing vibrations of different periods when it is agitated by strings that are themselves of different periods." But he is still bothered by the simultaneous propagation of different frequencies, a phenomenon which Mairan describes as an "incongruity" unexplained by Newton.

Two years after Cramer defended his thesis, a special chair in mathematics was created at the Geneva Academy for Cramer's part time occupancy. This left him with opportunities for travel. In May of 1727, Cramer went to Basel where he stayed for five months. He apparently impressed Johann Bernoulli with his culture and erudition. (Thirteen years later he was entrusted with the editing of both Johann and Jakob Bernoulli's complete works.) Presumably Bernoulli and Euler read his thesis or at any rate heard from him of its contents. As we shall see in Chapters 7 and 8, Cramer's visit occurred at just the time of Euler's and Bernoulli's early work in vibration theory. There is no reason to suppose that Cramer introduced them to the subject; but he certainly contributed to the general atmosphere of interest. He could not have made technical contributions; but one can imagine some connections between Euler's earliest work and Cramer's thesis: For example, Cramer describes the vibrations of a bell in terms of its cross sections as rings that vibrate in the shape of ellipses. It is just this description of a vibrating bell that motivated Euler in his analysis of the vibrating ring that we come to in the next chapter.[6]

---

[6] In the first paragraph of his paper on the vibrating ring (Chapter 7, section 1) Euler writes that his study will provide a basis for describing the oscillations of bells and it is clear from a marginal note that he considers the cross sections of the bell to vibrate as rings.

# 7. Euler (1727)

Leonhard Euler's very early attempt around 1727, while still in Basel, to treat the dynamics of the vibrating ring was not published until[1] 1862. Its early date was discovered by Truesdell.[2] Except for assuming the neutral surface to lie along an edge, Euler obtains the internal energy of an element of a naturally curved rod that is slightly deformed in a plane. Effectively, he defines and uses Young's modulus. However, he supposes an element of the vibrating rod to behave dynamically as a particle in one degree of freedom accelerated by a force given by a potential energy equal to the internal energy of the element. Euler analyses the vibrating circular ring by combining this dynamical notion with the assumption that the shape of the vibrating ring is elliptical. In section 7.1 we will first give a presentation of the vibrating ring that corrects Euler's presentation but otherwise deviates from it as little as we can manage. In this we will make parenthetic remarks on Euler's treatment. Then we will discuss Euler's own presentation. In section 7.2 we will discuss briefly two other early works of Euler on sound and vibration.

## 7.1. Vibrating Ring

We consider a two dimensional rod that is naturally curved along a circle. Let the width be $w$, Young's modulus be $E$ and the two dimensional density be $\rho$. The rod's configuration will be given by the curve of its central line $\mathbf{X}(s)$, where $s$ is arc length. Let $\mathbf{T} = \mathbf{X}'$ and $\mathbf{N} = \kappa^{-1}\mathbf{T}'$ denote the unit tangent and normal, where $\kappa$ is the curvature and differentiation with respect to $s$ is denoted by prime. Let $\mathbf{X}_0$, $\mathbf{T}_0$, $\mathbf{N}_0$, and $\kappa_0$ correspond to the rod in equilibrium and let the central line of the deformed rod be given by

$$\mathbf{X}(s) = \mathbf{X}_0(s) + u(s)\mathbf{N}_0(s) + v(s)\mathbf{T}_0(s). \tag{7.1}$$

---

[1] Euler [2].

[2] On the basis of the treatment of simple harmonic motion Truesdell conjectured that the paper belonged to Euler's Basel period. This was confirmed by Mikhailov on the basis of the handwriting and paper of the manuscript. (Truesdell [1], p. 143.)

We assume that the central line is the neutral line of the rod[3] so that $s$ is arc length for the curve $\mathbf{X}$ as well as $\mathbf{X}_0$. Since[4] $\mathbf{X}' = (1 - \kappa_0 u + v')\mathbf{T}_0 + (u' + \kappa_0 v)\mathbf{N}_0$, this means that $1 = \mathbf{X}' \cdot \mathbf{X}' = 1 - \kappa_0 u + v'$ to first order in the displacement. Thus, this condition is the requirement

$$v' = \kappa_0 u \tag{7.2}$$

in the small vibration limit. Now, $\kappa = |\mathbf{X}' \wedge \mathbf{X}''| = \kappa_0 + u'' + 3\kappa_0 v' - 2\kappa_0^2 u$, to first order (since $\kappa_0$ is constant), and with (7.2) this gives us that

$$\kappa - \kappa_0 = \kappa_0^2 u + u'' \tag{7.3}$$

in the small vibration limit. In terms of the change in curvature, we can calculate the tangential strain $\mu(s, x)$ at $\mathbf{X}(s) - x\mathbf{N}(s)$ or $(s, x)$ as we will denote the material point, $-w/2 \leqslant x \leqslant w/2$. Consider an element $ds$ of the rod. Then $ds$ is the distance from $(s, 0)$ to $(s + ds, 0)$; let $dl$ be the distance from $(s, x)$ to $(s + ds, x)$. Thus,

$$\mu(s, x) = \frac{dl - dl_0}{dl_0}.$$

But $ds$ is a differential arc of a circle of radius $\kappa^{-1}$ and $dl$ is the arc, subtending the same angle, of a circle of radius $\kappa^{-1} + x$; so $dl = (1 + \kappa x)\, ds$ and

$$\mu(s, x) = \frac{x(\kappa - \kappa_0)}{1 + \kappa_0 x} = x(\kappa - \kappa_0) \tag{7.4}$$

if $w \ll \kappa_0^{-1}$. The internal energy per unit (two dimensional) volume at $(s, x)$ is

$$\mathscr{E}(s, x) = \tfrac{1}{2}E\mu^2 = \tfrac{1}{2}Ex^2(\kappa - \kappa_0)^2 \tag{7.5}$$

and the internal energy per unit length is

$$\mathscr{E}(s) = \tfrac{1}{2}EI(\kappa - \kappa_0)^2 \tag{7.6}$$

where

$$I = \int_{-w/2}^{w/2} x^2\, dx = \frac{w^3}{12}. \tag{7.7}$$

(Euler, in assuming that the inside edge of the rod keeps its length, will have instead of $I$ the moment

$$I_E = \int_0^w x^2\, dx = 4I). \tag{7.8}$$

---

[3] This can be justified from the fact that the tangential force, $E \int \mu(s, x)\, dx$, in the notation to be given below, is small.

[4] See Chapter 3, note 6.

The total energy (which Euler won't use) is then $\int \mathcal{E}(s)\, ds$, the potential energy of deformation.

Now, in the case of the ring, $\mathbf{X}_0(s) = \kappa_0^{-1}(\cos \kappa_0 s, \sin \kappa_0 s)$, with $\kappa_0$ constant. We will use the variable $\Theta = \kappa_0 s$; so $f(\Theta) = u(s)$ and $g(\Theta) = v(s)$ are periodic with period $2\pi$ and $f = g'$. If we assume that the ring in equilibrium is unstrained, its potential energy in slight deformation is, from (7.6) and (7.3),

$$\tfrac{1}{2}EI\kappa_0^4 \int_0^{2\pi} (f + f'')^2 \, d\Theta. \tag{7.9}$$

Next, we impose the *pendulum condition* by requiring that the vibrating element $d\Theta$ be accelerated by the harmonic force

$$-\frac{\rho w}{a}(f\mathbf{N}_0 + g\mathbf{T}_0)\, d\Theta \tag{7.10}$$

or rather by requiring that the deformed ring be held in static equilibrium by the negative of this force. (This latter is in fact the form that Euler himself would introduce about eight years later (Chapter 10).) The total potential energy of the negative of the harmonic force (7.10) is

$$-\frac{1}{2}\frac{\rho w}{a}\int_0^{2\pi} (f^2 + g^2)\, d\Theta. \tag{7.11}$$

The condition of static equilibrium means that the variation of the sum of the energies (7.9) and (7.11) is zero:

$$\mathrm{var} \int_0^{2\pi} \left[ \tfrac{1}{2}EI\kappa_0^4 (g' + g''')^2 - \frac{1}{2}\frac{\rho w}{a}(g'^2 + g^2) \right] d\Theta = 0. \tag{7.12}$$

(This technique would not be available to Euler for another fifteen years or so.) This gives us the equation[5]

$$a\frac{EI\kappa_0^4}{\rho w}(g^{(6)} + 2g^{(4)} + g'') = (g'' - g) \tag{7.13}$$

where $g$ must have period $2\pi$. Up to a rotation and a factor of amplitude, $g$ must be of the form $g(\Theta) = \sin n\Theta$ with $n$ integral; whence (7.13) yields the corresponding period

$$2\pi\sqrt{a} = 2\pi\kappa_0^{-2}\left(\frac{\rho w}{EI}\frac{n^2 + 1}{n^2(n^2 - 1)^2}\right)^{1/2}. \tag{7.14}$$

---

[5] That is, replace $g$ by $g + \eta h$, with $h$ also periodic, differentiate with respect to $\eta$ and set this equal to zero at $\eta = 0$; then via partial integration one will have that the integral of $h[EI\kappa_0^4(g^{(6)} + 2g^{(4)} + g'') - (\rho w/a)(g - g'')]$ is zero and, since $h$ is arbitrary, that the part in [ ] is itself zero.

The case $n = 0$ corresponds to a simple translation of the ring with no restoring forces; the fundamental (the only case relevant for Euler's paper) is given by the case $n = 2$ for which the period is

$$2\pi\sqrt{\mathfrak{a}} = 2\pi\kappa_0^{-2}\frac{\sqrt{5}}{6}\sqrt{\rho w/EI} = 2\pi(w\kappa_0^2)^{-1}\frac{\sqrt{15}}{3}\sqrt{\rho/E} \qquad (7.15)$$

where we have used (7.7).

If the curve of the fundamental has a maximum normal displacement of $-\varepsilon$, its curve is given by $f = -\varepsilon \cos 2\Theta$, $g = -\frac{1}{2}\varepsilon \sin 2\Theta$, or

$$\kappa_0^{-1}(\cos\Theta, \sin\Theta) + \varepsilon \cos 2\Theta(\cos\Theta, \sin\Theta)$$
$$-\tfrac{1}{2}\varepsilon \sin 2\Theta(-\sin\Theta, \cos\Theta), \qquad (7.16)$$

which has the same curvature as a function of $\Theta$ as does the ellipse

$$([\kappa_0^{-1} + \varepsilon]\cos\Theta, [\kappa_0^{-1} - \varepsilon]\sin\Theta) \qquad (7.17)$$

to first order in $\varepsilon$. In fact, the curvature of (7.16), given by (7.3) to first order in $\varepsilon$, is

$$\kappa = \kappa_0 + 3\kappa_0^2\varepsilon \cos 2\Theta. \qquad (7.18)$$

The curvature of the ellipse (7.17) is $(\kappa_0^{-1} + \varepsilon)(\kappa_0^{-1} - \varepsilon)/[(\kappa_0^{-1} + \varepsilon)^2 \cos^2\Theta + (\kappa_0^{-1} - \varepsilon)^2 \sin^2\Theta]^{3/2}$, which is also equal to (7.18) to lowest order. But the curves are different even in lowest order since $\kappa_0^{-1}\Theta$ is not arc length for the ellipse. (Euler will not derive a shape curve for the vibrating ring; he will simply assume it to be the ellipse, with semi-axes $(\kappa_0^{-1} \pm \varepsilon)$, whose curve is given by (7.17).)

We come now to Euler's work. He assumes that an edge of the ring keeps its length under deformation; this changes what we did above in that $x \in (0, w)$ rather than the interval $(-w/2, w/2)$. Except for the fact that he singles out $dl - dl_0 = \mu(s, x)\,ds$ rather than the strain $\mu$ itself, Euler derives the equality (7.4) as we did above.[6] To find the force which this change of length will cause on the sectional element $dx$, Euler proceeds as follows: He supposes that a (two dimensional) rod of width $f$ and length $dl_0$ requires a force $P$ to draw it to a length $dl$ and in general that the required force is proportional to the width and the change of length.[7] In effect then, he defines Young's modulus $E = P\,dl_0/f(dl - dl_0)$, although without using a simple notation for it and without noticing—or at any rate, pointing out—the general relevance of this constant for a material. Still, he has now that

---

[6] Euler considers the change in length of the outside edge of an element of the ring. In his notation: $E\varepsilon$ (our $dl - dl_0$) is given by $((a - b)c\,ds)/ab$, where $a$ is our $\kappa_0^{-1}$, $b$ is our $\kappa^{-1}$ (so that $(a - b)/ab$ is our $\kappa - \kappa_0$), and $c$ is our $w$.

[7] In Euler's terminology: "Let [the] cohesion of the parts or filaments be of such a force that a row [of filaments] $FG = f$ extended to $FJ = g$ is able to sustain a weight $P$." That $g$ refers to the change in the length of the filaments is clear from Euler's use of it.

the force on the sectional element $dx$ is[8] $Ex(\kappa - \kappa_0)\,dx$. The force $(1/w)Ex^2(\kappa - \kappa_0)\,dx$ applied at the edge sectional element $dx$ at $(s, w)$ produces the same torque about $(s, 0)$ as does the force $Ex(\kappa - \kappa_0)\,dx$ applied at $(s, x)$; so by integrating, Euler obtains the total force to be applied on the edge sectional element $dx$ at $(s, w)$ to produce the appropriate torque about $(s, 0)$. This total force is[9] $(1/w)EI_E(\kappa - \kappa_0)$, where $I_E$ is given in (7.8). In a deformation that changes the curvature an amount $d\kappa$, the change in $dl$, at $x = w$, given by (7.4), is[10] $w\,d\kappa\,ds$. Thus, Euler has that the work done in the deformation of the ring element $ds$ is[11]

$$\frac{1}{w} EI_E(\kappa - \kappa_0)w\,d\kappa\,ds = \frac{d\mathscr{E}}{d\kappa} d\kappa\,ds, \qquad (7.19)$$

where the right side gives the energy notation of (7.6) with $I_E$ replacing $I$.

Euler assumes that the ring vibrates in the shape of the ellipse that we parametrized in (7.17). He concerns himself only with a point of maximum curvature since the half period for the vibration of this point is the half period for the ring. This point is given by $\Theta = 0$ (or $\pi$) in (7.17) or (7.16) and its curvature, given by (7.18), is[12]

$$\kappa = \kappa_0 + 3\kappa_0^2 \varepsilon. \qquad (7.20)$$

Corresponding to the change $d\varepsilon$, the work of deformation of the ring element at this point of maximum curvature is by (7.20) and (7.19)

$$\frac{d\mathscr{E}}{d\kappa}\frac{d\kappa}{d\varepsilon} d\varepsilon\,ds = 3\kappa_0^2 \frac{d\mathscr{E}}{d\kappa} d\varepsilon\,ds = 9\kappa_0^4 EI_{E\varepsilon} d\varepsilon\,ds \qquad (7.21)$$

---

[8] In Euler's notation: This is the force "drawing together the sides Ee and $\varepsilon e$" along the section $dx$, which he writes as $(Px\,dx\,dt/cfq)$, where $dt$ is the arc E$\varepsilon$, given in note 6. Since $g$ is our $dl - dl_0$ (and $ds$ can be replaced by $dl_0$ in our notation) Euler's $(P\,ds/fg)$ is our $E$.

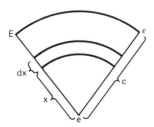

[9] In Euler's notation: the total force is $(Pc^2(a - b)\,ds/3abfg)$.

[10] Euler actually writes this explicitly only for the case of an extremal point of an ellipse (see note 13).

[11] Euler does not refer to this quantity as work, but it is clear that he considers it as analogous to $F\,dy$ in the case of one degree of freedom.

[12] Euler actually uses the exact expression for the maximum curvature of the ellipse, $(a + \omega/(a - \omega)^2)$, where $\omega$ is our $\varepsilon$.

as Euler finds (except that he makes an arithmetical error that drops the factor of three).[13] This is (otherwise) wrong only in that $I_E$ stands in the place of $I$.

Now Euler makes an assumption about the dynamics of the ring element at the point of maximum curvature. He assumes with mathematical neatness and physical naiveté that it is accelerated by a force $Q$ such that the work, $Q d\varepsilon$, done in moving this element an amount $d\varepsilon$ is equal to the work given in (7.21):

$$Q = 9\kappa_0^4 EI_E \varepsilon \, ds. \tag{7.22}$$

But Euler has found a force $Q$ on the mass $\rho w \, ds$ that depends linearly on the displacement $-\varepsilon$ from its equilibrium position. He recognizes this as analogous to the simple pendulum and finds thus the corresponding period[14] (see (1.2) and (1.3))

$$2\pi\kappa_0^{-2}\tfrac{1}{3}\sqrt{\rho w / EI_E} = 2\pi(w\kappa_0^2)^{-1}\frac{1}{\sqrt{3}}\sqrt{\rho / E} \tag{7.23}$$

(except that he doesn't have the factor $1/\sqrt{3}$ because of the arithmetical error that he has in (7.21).)

---

[13] Euler seeks the product of the contracting force and the change in $E\varepsilon$ associated with a displacement $dz$ towards the center. He finds the change in $E\varepsilon = ((a-b)c \, ds/ab)$ by tediously subtracting the value with $((a-\omega)^2/(a+\omega))$ in the place of $b$ from the value with $((a-\omega+dz)^2/(a+\omega-dz))$ in the place of $b$. He obtains $((a+3\omega)/(a-\omega)^3)c \, dz \, ds$, but as one can see immediately by differentiating $((a+\omega)/(a-\omega)^2)$, the result should be $(3a+\omega/(a-\omega)^3)c \, dz \, ds$. The product of this with the contracting force given in note 9 with $((a-\omega)^2/(a+\omega))$ in the place of $b$, is, for small deformations, $([3]Pc^3 w \, ds^2 \, dz/a^4 fg)$, which Euler calls $Q \, dz$. (In our notation $Q$ is given in (7.22).)

[14] In recognizing the *length of the simple isochronous pendulum* in the force $Q$, Euler displays an archaic and bizarre perspective on the *momentum principle* in the case of one degree of freedom for conservative systems: Such systems are described by conservation of energy, $U + \tfrac{1}{2}m\dot{y}^2 = \text{const.}$; let $h$ be the height from which a body should fall to acquire the velocity $\dot{y}$, so $\tfrac{1}{2}m\dot{y}^2 = \mathfrak{g}mh$; make this substitution, differentiate, and substitute the force $F = -dU/dy$ to obtain the equation

$$dy/\mathfrak{g}m = dh/F$$

This is the form of the *momentum principle* that Euler seems to favor. Now, in the case of an *harmonic force*, $-\mathfrak{a}^{-1}my$, this equation can be written

$$dh = -\frac{1}{\mathfrak{a}\mathfrak{g}} y \, dy.$$

Presumably from having worked out the simple pendulum problem using this equation, Euler recognizes at this point that $\mathfrak{a}\mathfrak{g}$ is the *length of the simple isochronous pendulum*. Euler writes both these equations for the case at hand. In his notation: The weight $\mathfrak{g}m$ is written $Ac \, ds/e^3$; the force is $Q$, given in the previous note; $y$ is $\omega$; and $h$ is $\nu$. Then the length $\mathfrak{a}\mathfrak{g}$ in the second equation becomes $Aa^4 fg/[3]Pc^2 e^3 \, ds$ which in our notation is $\mathfrak{g}\rho w/9\kappa_0^4 EI_E$. It is only after obtaining this result that Euler recognizes that the $ds$ is part of the definition of what we have called Young's modulus, $E$, and so does not cause him any problem as an inappropriate infinitesimal.

To summarize: Euler's paper is obviously an early work which he didn't intend to publish. Compared with papers that he wrote only a few years later, it is circuitous in its handling of differential calculus and it lacks clarity. Considering how long it took Young's modulus to become understood, one certainly appreciates the paper. But it has four errors: there is the assumption (by which he avoids Taylor's main problem) that the vibrating shape is an ellipse; but this doesn't lead to an error in the internal energy of the (material) elements at the points of extremal curvature. There is the arithmetical error arising in his circuitous calculus. There is the assumption that the edge keeps its length, which leads to the error of $I_E$ in the place of $I$. Finally, there is the assumption in (7.22) that the work done in accelerating the element is the change in its internal energy.[15] It is an idea that is taken too directly from dynamics in one degree of freedom. It is an assumption of *local* conservation of energy!

## 7.2. Sound

In 1727 Euler wrote a short dissertation *De Sono*[16] on the production and propagation of sound. This work is somewhat expanded in the first chapter of his book *Tentamen . . . Musicae*[17] which he wrote about four years later though it was published only in 1739. Were it not that these works were written by Euler himself, they would be unremarkable. They collect various vaguely worded qualitative assertions about the nature of sound which fall short of the knowledge that was left by the seventeenth century; in addition they contain a formula for the velocity of sound that is Newton's result multiplied by $4/\pi$ and a formula for the vibrational frequency of a musical string that is Taylor's result; but they give no hint of his derivation of either.

In these works one can admire Euler's impish disrespect for authority, knowing of course that it is Euler himself who will later do more than anyone else to obtain an analytic understanding of the subject. We will discuss these works briefly:

Euler conceives of air as made up of infinitely small but nevertheless springy particles that are compressed together. One can presume that this picture entails Boyle's law. Euler has the qualitative picture of sound as

---

[15] Note, for what it's worth, that only this last mistake enters the result for the period if $I_E$ is replaced by $I = \frac{1}{4}I_E$ in (7.23) which then becomes $2\pi(w\kappa_0^2)^{-1}\sqrt{\frac{4}{3}}\sqrt{\rho/E}$ or 0.89 times the correct period given in (7.15). With $I_E$ itself, this factor becomes 0.45; with the arithmetical error, 0.78.

[16] Euler [3].

[17] Euler [7]. In the quotations we have used the translation by Smith. The other chapters of this book deal with speculative music; for a discussion see Dostrovsky and Cannon [1].

being a pressure wave with pitch described by frequency in the case of a "simple" or periodic sound. He sees a composite sound as made up of simple sounds, sounding consonant when the frequencies bear a simple ratio. Supposedly on the basis of this picture, Euler has calculated the velocity of sound; but he gives only the result[18]

$$V = \frac{4}{\pi} \sqrt{P/\rho} \qquad (7.24)$$

where $P$ is the pressure and $\rho$ the density. In Euler's notation this is given as $4\sqrt{3166nk}$ where 3166 (scruples) is the length of a "second's pendulum" having a period of 2 seconds so that $1 = \pi\sqrt{3166/g}$, $n$ is the specific density of mercury in air, and $k$ is the height of a mercury barometer: $n = \hat{\rho}/\rho$ and $k = P/\hat{\rho}g$. Euler is very pleased to have a result that is larger than Newton's; he manages to ignore the fact that it is too large. (In a letter of 1737 to Johann Bernoulli,[19] Euler still stood by his derivation). He says in his very first paragraph of the dissertation that Newton met with little success in explaining the true nature of sound propagation. But whatever Euler's own model of sound propagation may have been at this time, he would later abandon it to make a careful study of Newton's.[20]

Euler gives the vibrational frequency of a musical string. In his notation, he gives the double frequency as $(22/7)\sqrt{3166P/aq}$ where $\frac{22}{7}$ is an approximation for $\pi$, $P$ is the tension, $a$ the length, and $q$ the total weight of the string.[21] This agrees with Taylor's result as well as Johann Bernoulli's, as Euler says in the dissertation though he gives no reference in the book. Again we have no idea what Euler's derivation may have been. Truesdell mentions seeing a note from about this time in which Euler derives a frequency that is wrong by a factor $2/\sqrt{\pi}$, corrected in the hand of Johann Bernoulli.[22] We have looked at some other early notes of Euler,[23] also pointed out by Truesdell, in which a derivation is attempted. We have not made much of these notes. They contain no physics; and we have the

---

[18] In the book Euler gives only a numerical result (1100 Rh.ft./sec.) (par. 6).

[19] Eneström [1], V, p. 256.

[20] Truesdell [2] asserts that Euler makes an "essential contribution" in being the "first" to derive the velocity of sound on the basis of the "Townley–Boyle law" (p. XXVI). We find this assertion to be too strong. Euler does not present his derivation and it is wrong anyway. Hermann had given a wrong derivation on the basis of Boyle's law ten years before and Newton, of course, had given a correct derivation forty years before. For Euler's later study of Newton's work, see Truesdell [2], p. XXXII. It might be of interest if Euler's early derivation were found among his papers.

[21] In the book he gives the same result with a better approximation for $\pi$ and slightly different notation: $\frac{355}{113}\sqrt{3166n/a}$, where $n$ is $p/q$ (par. 9).

[22] Truesdell [1], p. 142.

[23] Euler [1], pp. 133–140, 146–147. We would like to thank Prof. Walter Habicht and Beatrice Bosshart of the University of Basel for photographs of these pages.

impression that Euler was hoping to grasp the answer as a nice geometrical relationship.

Euler measured the lengths, weights, and tensions of two strings of known musical pitch and calculated their absolute frequencies.[24] He does not say anything about experimental confirmation.

Euler also gives a formula for the double frequency of vibrations in a cylindrical pipe open at both ends,[25] namely $l^{-1}\sqrt{P/\rho}$. He characterizes these vibrations as longitudinal pressure waves; however he obtains the frequency not by an analysis of the pressure wave but rather by making a direct analogy between the pipe and the string: In the place of the tension of the string he puts the pressure ($P$) of the air since it provides the restoring force; and in the place of the weight of the string he puts the weight ($g\rho l$) of the air in the pipe. Were Euler taking the half wavelength of sound to be the length of the pipe, his formula for the frequency would correspond to Newton's velocity and not his own. Thus he seems not to identify the wavelength in the pipe with the wavelength of propagating sound.

Euler mentions higher vibrational modes only in his book, motivated, it seems, by the problem of understanding why "the [blowing] pressure can be so great that a pipe will produce a sound an octave higher than its normal sound, and a further increase in pressure will produce the twelfth then the fifteenth, etc."[26] He writes of a "fixed limit" to the blowing pressure for a given pitch which he explains as follows: as the blowing pressure increases, the amplitude in the pipe increases until it becomes too large for the diameter of the pipe, at which point a node is introduced. Referring to Sauveur, he describes the phenomenology of the higher modes of the vibrating string.

Euler remarks that large vibrations do not have the properties of iso-chronism and simultaneous crossing of the axis: "In the case of too great initial amplitude, the larger vibrations are not isochronous with the smaller, whence the sound gradually becomes lower and does not remain the same. Then it easily happens that the whole string does not complete all vibrations at the same time, but one part may reach its maximum speed or state of rest more rapidly or more slowly than another part, so that the sound is uneven and harsh."[27] Presumably on the supposition that a slack string requires a greater amplitude to produce a given volume, he writes that "if the string does not have sufficient tension, the sound will be less pleasing because the irregularities of vibrations will be too great and will simply

---

[24] On the basis of these measurements he reports in the dissertation that a $D\#$ has a frequency of 559 semi vibrations per second, and, in the book, that an $A$ has a frequency of 392 semi vibrations per second (par. 10).

[25] In Euler's notation: $\frac{22}{7a}\sqrt{3166nk}$, where $a$ is our $l$. In the book he gives the same result with, again, the better approximation for $\pi$ and slightly different notation (par. 34).

[26] par. 40.

[27] par. 20; there are similar remarks in pars. 10, 16, and 17.

stir the air rather than effect regular oscillations." For this reason he recommends that musical strings be as tense as possible. (When Daniel Bernoulli finally saw Euler's book, about ten years after it was written, he remarked that there would be departures from isochronism also for strings that are so tense as to be near rupture, "for the elongation will not be proportional to the extending force ... and everything must be irregular."[28])

---

[28] Letter of 28 January 1741, in Fuss [1], II, p. 470.

# 8. Johann Bernoulli (1728)

Johann Bernoulli announced his results on the vibrating string in letters of October and December of 1727 to his son Daniel in St. Petersburg. Excerpts of these letters were published by the St. Petersburg Academy in its volume for 1727 which appeared[1] in 1729. He presented the Academy with details in a paper that was published in the volume for 1728 which appeared[2] in 1732. To a considerable extent this paper can be regarded as a set of notes on Taylor's analysis. But in addition, it introduces an analysis of the corresponding discrete problem of the weightless string loaded with equal and equally spaced point masses; it gives the *pendulum condition* for the continuous case in the form $â \, d^2y/dz^2 = -y$ that was circumvented by Taylor; and finally it introduces the potential and kinetic energies as the sums or integrals of the respective energies of the masses or string elements. In solving the equation of the pendulum condition Bernoulli suppresses the *harmonic constant* $a$; he then finds this constant in two ways—either by substituting the solution back into the pendulum condition or by finding the maximum velocity of the mid-point from the equality of the maximum potential energy with the maximum kinetic energy.

## 8.1. Vibrating String (Continuous and Discrete)

As had Taylor, Bernoulli makes the assumptions of isochronism and simultaneous crossing of the axis, considering only the case of a string released from one side; on the basis of these assumptions he determines that the pendulum condition must be satisfied. These are the central ideas in dynamics at the time. They were introduced by Taylor but given prominence by Bernoulli who writes in the first sentence of his paper that "the vibrating string must compose itself into such a shape that all the weights reach the axis $AB$ simultaneously, whence it follows that the [rates of increase of the] velocities, and thus also the accelerating forces must be proportional to the distances to be traversed." (He does not mention

---

[1] Johann Bernoulli [3].
[2] Johann Bernoulli [4].

Taylor's paradox or anything about how the string attains the appropriate shape.)

We will continue to use the notation that we used in discussing Taylor's analysis. Thus the string is stretched along the $z$-axis in the $z$–$y$ plane from $z = 0$ to $z = l$, with tension $P$ and linear mass density $\rho$. In the case of the loaded string, $n$ point particles are placed along a massless string at equal distances $\Delta z = l/(n + 1)$, each particle having mass $((n + 1)/n)\rho\,\Delta z$; their locations, in the case of small vibrations, are $(z_i, y_i)$, $i = 1, 2, \ldots, n$, and the end points are $(z_0, y_0) = (0, 0)$ and $(z_{n+1}, y_{n+1}) = (l, 0)$. In the continuous case, let $\alpha$ again denote the angle between the tangent and the $z$-axis; in the discrete case, let $\alpha_i$ denote the angle between the line through $(z_{i-1}, y_{i-1})$ and $(z_i, y_i)$ and the $z$-axis. Write $\Delta\alpha_i = \alpha_{i+1} - \alpha_i$.

From "static principles", Bernoulli finds by elementary geometry that the force[3] on the $i$th particle is $2P \cos((\pi - \Delta\alpha_i)/2) = 2P \sin(\Delta\alpha_i/2)$ or, in the small vibration limit, $P\Delta\alpha_i$. In the continuous case, he has by analogy that the force on the mass element $\rho\,dz$ is[4] $P\,d\alpha$. (Thus, he has this force only in the small vibration limit and not in general as had Taylor.) Since he is working in the small vibration limit, the forces are perpendicular to the $z$-axis and he can write them in the forms[5]

$$P\Delta\alpha_i = P\frac{y_{i+1} - 2y_i + y_{i-1}}{\Delta z} \quad \text{and}^6 \quad P\,d\alpha = P\frac{d^2y}{dz^2}\,dz. \qquad (8.1)$$

(This limiting form for the curvature, $d^2y/dz^2$, was circumvented by Taylor who attained the limit after considering the equation (3.2).)

---

[3] In Bernoulli's terminology: "the tension of the string is to the force by which any weight, for example E [see figure] is urged towards e as the sine of the angle DEe is to the sine of the angle DEF or IEF, that is (because the string is almost straight and the weights are equally spaced), as the whole sine [DE or de, our $\Delta z$] is to FI."

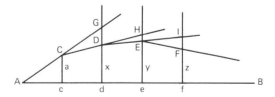

[4] In Bernoulli's terminology: the force on a "particle KS of the string" can be taken as $P(FG/KG)$ where G and F are the intersection points of the perpendicular bisector of the string with the tangents at K and S, respectively (prob. VII). If FG is infinitesimal and the small vibration limit is taken, (FG/KG) is $d\alpha$ in our notation.

[5] For example, Bernoulli writes that FI (our $-\Delta\alpha_3\Delta z$) is equal to $2y - x - z$ (our $2y_3 - y_2 - y_4$).

[6] Bernoulli writes (prob. VII) the general expression for the "radius of osculation" of a point on the string, and he remarks that since the string is "exceedingly elongated" the radius of osculation can be approximated by $dy^2/d\,dx$, i.e. $(-d^2y/dz^2)^{-1}$ in our notation, since $y$ is our $l/2 - z$ and $x$ is our $c - y$.

Bernoulli can now write the pendulum condition for the respective cases as the equations[7]

$$\hat{a}\,\frac{y_{i+1}-2y_i+y_{i-1}}{(\Delta z)^2}=-y_i \quad \text{where } \hat{a}=anP/(n+1)\rho \tag{8.2}$$

and[8]

$$\hat{a}\,\frac{d^2 y}{dz^2}=-y \quad \text{where } \hat{a}=aP/\rho. \tag{8.3}$$

He integrates[9] the latter, first to the form $\hat{a}\,dy^2=(c^2-y^2)\,dz^2$, and then by familiar geometric methods to find, as we write functionally, that $y = c\sin(z/\sqrt{\hat{a}})$ with $\pi^2\hat{a}=l^2$. (The boundary conditions are dealt with implicitly in the *geometric* approach which also seems to hide the higher mode possibilities, $N^2\pi^2\hat{a}=l^2$.) Thus, he demonstrates "that the vibrating string . . . assumes the shape of an elongated companion of the trochoid." In the discrete case, he treats the system of equations (8.2), $i = 1, 2, \ldots, n$, by elementary algebraic manipulations[10] for $n \leqslant 7$. He does not obtain the solutions in a form that would be analogous to the continuous case, e.g. $y_i = c\sin\Omega i$ with $\Omega = \pi/(n+1)$ and $1-\cos\Omega=(\Delta z)^2/2\hat{a}$; but we will use this notation for convenience.

Now Bernoulli has solved (8.2) and (8.3) for the shape curves and also for the constants $\hat{a}$, except that he loses the constants in the solutions. Therefore he has the problem of rediscovering the harmonic constants. He does this in two ways. One way is to substitute the solutions $y_1 = c\sin\Omega$ and $y = c\sin(\pi z/l)$ back into (8.2) and (8.3) respectively. Whence, Bernoulli finds the harmonic constants:

$$a = \frac{l^2}{2n(n+1)(1-\cos\Omega)}\frac{\rho}{P} \tag{8.4}$$

---

[7] In Bernoulli's notation: $2a - x : a = 2x - a - y : x = 2y - x - z : y = 2z - y - t : z$, etc., where $a, x, y, t$ are our $y_i$, $i = 1, 2, \ldots$, etc.

[8] In Bernoulli's notation: $(FG/KG)Mg = f KH\, dL$, where $f$ is our $a^{-1}$, $KH$ is our $y$, and $dL$ is our $\rho\,dz$; (see notes 5 & 7 above for details about the left hand side of the equation).

[9] The equation is integrated in the concluding paragraph of problem VII, where it is written in the form $n^2a^2d\,dx : dy^2 = a - x$, where $x$ and $y$ are defined in our notation in note 7 above, $a$ is our $c$ (so $a - x$ is our $y$), and $n^2$ is our $\hat{a}/c^2$. Bernoulli writes: "Multiplying both sides of the equation by $dx$, one will have $n^2a^2\,dx\,ddx : dy^2 = a\,dx - x\,dx$; and, taking the integrals [using $dx\,ddx = \tfrac{1}{2}d(dx)^2$], $n^2a^2\,dx^2 = (2ax - x^2)dy^2$, whence $dy = na\,dx : \sqrt{2ax - x^2}$."

[10] For example, for $n = 3$: "$y = a$, $z = 0$, . . . thus $2a - x : a = 2x - 2a : x$, whence $2ax - x^2 = 2ax - 2a^2$, & $x = a\sqrt{2}$" (in the notation given in note 8 above).

in the discrete case[11] and, in the case of the vibrating string, he finds Taylor's result[12] (3.5). He determines the corresponding periods,[13] (1.3), by integrating $\mathfrak{a}v\,dv = -y\,dy$, as had Hermann[14] (section 5.3).

Otherwise, he re-obtains the harmonic constants using the equality (1.5), $\mathfrak{a}V^2 = c^2$. That is, for a given amplitude $c$, he seeks the maximum velocity $V$ of the mid-point of the string and in the discrete case, given the amplitude of the first mass, he seeks its maximum velocity. To do this, he introduces the maximum kinetic and potential energies of the vibrating string and the loaded string. This entails a lot of work simply to find a suppressed constant; but it is remarkable enough if only because it introduces kinetic energy as well as potential energy for a system in many degrees of freedom. We will discuss it in the remainder of this section.

Bernoulli defines the potential energy of the string as $P$ times the string's change in length from its equilibrium length, namely

$$P\left[\sum_{i=0}^{n} \Delta z\left(1+\left(\frac{y_{i+1}-y_i}{\Delta z}\right)^2\right)^{1/2} - l\right] = P\sum_{i=0}^{n}\frac{1}{2}\left(\frac{y_{i+1}-y_i}{\Delta z}\right)^2 \Delta z \qquad (8.5)$$

in the discrete case of the loaded string and

$$P\left[\int_0^l\left(1+\left(\frac{dy}{dz}\right)^2\right)^{1/2} dz - l\right] = P\int_0^l\frac{1}{2}\left(\frac{dy}{dz}\right)^2 dz \qquad (8.6)$$

in the continuous case, where the equalities hold in the small vibration limit. He does this by supposing the tension to be due to a weight $P$, that hangs in the gravitational field, which is lifted when the string is deflected, the string being of inextensible material. He refers only to the *vis viva* of the weight $P$ and not to the potential energy of the string. If one substitutes $y_i = c\sin\Omega i$ into (8.5), one finds that the sum equals

$$P\frac{c^2}{\Delta z}\left[\sum_{i=1}^{n}\sin^2\Omega i - \sum_{i=1}^{n}\sin\Omega i\sin\Omega(i+1)\right]$$

$$= P\frac{c^2}{\Delta z}(1-\cos\Omega)\sum_{i=1}^{n}\sin^2\Omega i \qquad (8.7)$$

---

[11] Bernoulli needs to consider (8.2) for $i = 1$ only. In his notation this is $(2a - x)Mg : b = fa(1/n)L$, where $Mg$ is our $P$, $b$ is our $\Delta z$, $f$ is our $\mathfrak{a}^{-1}$, and $L$ is our $(n+1)\rho\Delta z$.

[12] In his notation: Bernoulli has that $\mathsf{FG/KG}$ $[= (d\,dx/dy^2)dy = ((a-x)/a^2n^2)dy] = (a-x)dx/na\sqrt{2ax-x^2}$ and that $dL$ (which is $\rho\,dz$ in our notation) $[= (L/AB)dy] = na\,dx\,L/AB\sqrt{2ax-x^2}$ (see note 10). He substitutes these into the *pendulum condition* given in note 9 to obtain $f = ABMg/Ln^2a^2$ and he has that $n^2 = l^2/p^2a^2$; so $f = p^2Mg/ABL$, where $p$ is $\pi$, $\mathsf{AB}$ is our $l$, and $L$ is our $\rho l$.

[13] In Bernoulli's notation: the quarter period is $p:2\sqrt{f}$. So having obtained $f$ (note 12), Bernoulli writes the quarter period for the string as $\sqrt{ABL}/2\sqrt{Mg}$ and the frequency as $p\sqrt{DP}/\sqrt{ABL}$, "just as Taylor found," where $D$ is the length of a standard pendulum, and in the last expression, $L$ is the weight of the string rather than its mass.

[14] Truesdell [1] suggests that Bernoulli was the first to treat simple harmonic motion by straightforward integration of the differential equation (p. 134n). However, Hermann had done this a dozen years before.

Bernoulli calculates this in his form[15] for the cases $n \leq 6$. He also carries out the integration in (8.6) for $y = c \sin (\pi z/l)$, obtaining[16]

$$\frac{\pi^2}{4l} Pc^2. \tag{8.8}$$

With $c$ the amplitude of the vibrations, (8.7) and (8.8) give the maximum potential energies.

Bernoulli defines the kinetic energies to be the sum of the kinetic energies of the mass particles in the discrete case and the integral of the differential kinetic energies of the string elements in the continuous case. Actually, he considers these energies only at their maximum values which occur when the string passes through the axis. Let $v_i$ denote the maximum velocity of the $i$th mass, in the discrete case, and let $v(s)$ denote the maximum velocity of the string element $\rho\, dz$, in the continuous case. The maximum kinetic energies are then respectively

$$\sum_{i=1}^{n} \frac{1}{2} \frac{n+1}{n} \rho \Delta z v_i^2 \quad \text{and} \quad \int_0^l \tfrac{1}{2}\rho v^2 \, dz. \tag{8.9}$$

Now, Bernoulli uses the fact that in simple harmonic motion the maximum velocity is proportional to the maximum displacement (see (1.5)). Therefore, using the solutions to (8.2) and (8.3), he has that

$$v_i = V \sin \Omega i \quad \text{and} \quad v(z) = V \sin \frac{\pi z}{l} \tag{8.10}$$

respectively, where $V$ is the maximum velocity of the mid-point of the string in the continuous case and, $v_1/\sin \Omega$, in the discrete case. Bernoulli puts (8.10) into (8.9), obtaining the maximum kinetic energies[17]

$$\frac{1}{2} \frac{n+1}{n} \rho \Delta z V^2 \sum_{i=1}^{n} \sin^2 \Omega i \tag{8.11}$$

---

[15] For example, for $n = 3$, Bernoulli writes $(4a^2 - 2a^2\sqrt{2})P : b$, where $a$ is our $c/\sqrt{2}$ and $b$ is our $\Delta z$.

[16] In Bernoulli's notation: the change in length of an element of the string is $ds - dy$, where $-s$ is arc length and $dy$ is our $-dz$. Bernoulli approximates this by $dx^2/2dy$, where $x$ is our $c - y$, and finds the change in length for half the string, $\int dx \sqrt{2ax - x^2}/2na = AB/8n^2$, where $AB$ is our $l$, $a$ is our $c$, and $n$ is our $l/\pi c$, so that the "living force of the weight $P$" is $PAB/4n^2$ (prob. VII).

[17] For example, for $n = 3$, the *vires vivae* of the three particles is $\frac{4}{3}zL$, where $z$ is "the height through which a body descending freely acquires a velocity equal to that which the [first] particle has when it reaches [the axis]". Bernoulli gives the velocity of the first (and third) particles as $\sqrt{z}$ and that of the second particle as $\sqrt{2z}$, (see note 10), although the velocities should be $\sqrt{2z}$ and $\sqrt{4z}$, respectively.

in the discrete case and[18]

$$\frac{\rho l}{4} V^2 \tag{8.12}$$

in the continuous case. As a matter of conservation of energy, he has that (8.7) and (8.11) are equal as are (8.8) and (8.12). He knows as we said above that the harmonic constant $\mathfrak{a}$ is $c^2/V^2$ ($y_1^2/v_1^2$, in the discrete case); so he solves these equalities for $c^2/V^2$ obtaining the results (8.4) and (3.5) (except that he makes an error of a factor of 2 that arises in his definition of kinetic energy).

## 8.2. Remark on the Energy Method

Bernoulli's energy method can be separated from the solution for the shape of the vibrating string. If the pendulum condition is used only in the form $\mathfrak{a}v_{max}^2 = y_{max}^2$, (1.5), then the equality of the maximum potential energy and the maximum kinetic energy, from (8.6) and (8.9), in the continuous case, gives

$$\mathfrak{a}P \int_0^l \left(\frac{dy_{max}}{dz}\right)^2 dz = \rho \int_0^l y_{max}^2 dz. \tag{8.13}$$

Effectively then, Bernoulli substitutes $y_{max} = \sin(\pi z/l)$ and solves for $\mathfrak{a}$.

It seems that equation (8.13) was indicated already by Huygens, in his personal notes on the vibrating string[19] of 1673. Huygens did not carry through any details. He indicates however, that since the vibrations are small, $y_{max}$ can be approximated quadratically: $y_{max} = l^2/4 - (z - (l/2))^2$. If this is substituted into (8.13), one obtains Sauveur's result (4.4) for $\mathfrak{a}$. It is possible that equation (8.13) was known since Huygens' time.

---

[18] In Bernoulli's notation: the velocity of a "point" of the string when it reaches the axis is $((a-x)/a)\sqrt{S}$, where $S$ is the height through which a body must fall to acquire the maximum velocity of the mid-point of the string. Again there is an error of $\sqrt{2}$. He obtains the total *vis viva* $\frac{1}{2}LS$, which is twice the total kinetic energy.

[19] Huygens [1], XVIII, pp. 489–495. The editors have carried through the details of Huygens' method to obtain the vibrational frequency.

# 9. Daniel Bernoulli (1733; 1734); Euler (1736)

Daniel Bernoulli and Euler discussed the problem of the small oscillations of a hanging chain or its discrete analogue, the linked pendulum, in St. Petersburg before Bernoulli's departure in 1733. Bernoulli submitted his results[1] to the St. Petersburg Academy before leaving and communicated his proofs[2] the following year. After receiving Bernoulli's proofs, Euler submitted his own version to the Academy.[3] These papers finally appeared in 1738, 1740, and 1741, respectively. Essentially, these works treat the hanging chain and the linked pendulum as Johann Bernoulli had treated the vibrating string and the loaded string; but where Taylor and Johann Bernoulli had found the shape of the vibrating string to be sinusoidal, these works find the corresponding shape to be given by a Bessel function or, in the discrete case, by Laguerre polynomials, thus introducing these functions and the problem of finding their zeros. Perhaps it is because these functions are defined only analytically and not geometrically (as was the sine function) that the higher zeros are not overlooked and higher modes are discovered mathematically.

We will discuss these works together, pointing out respective contributions along the way. Facsimile of Bernoulli's papers and translations are given in the appendix.

## 9.1. Linked Pendulum and Hanging Chain

Bernoulli and Euler restrict themselves from the start to the case of small oscillations and the case of isochronism and simultaneous crossing of the axis and follow Taylor and Johann Bernoulli in assuming as a consequence that the *pendulum condition* holds. They treat the discrete case first and obtain the continuous case as a limiting analogue, as Johann Bernoulli had done.

We will continue to use the notation that we used in describing Johann Bernoulli's analysis. Thus the chain hangs (in a gravitational field, $(-g, 0)$)

---

[1] Daniel Bernoulli [4].

[2] Daniel Bernoulli [5].

[3] Euler [6].

in the $z$–$y$ plane along the $z$-axis from the point $(l, 0)$ with its free end at the origin, when in equilibrium. When a small displacement is made, the chain will describe a curve $y(z)$, $0 \leqslant z \leqslant l$ and $\alpha(z)$ will denote the angle between the tangent and the $z$-axis. In the case of the linked pendulum, masses $m_i$ are hung along a massless string so that, in the case of small vibrations, their locations are denoted $(z_i, y_i)$, $i = 0, 1, 2, \ldots n-1$, with the $z_i$ fixed, $z_0 = 0$ and $(z_n, y_n) = (l, 0)$. Let $\Delta z_i = z_{i+1} - z_i > 0$ denote the length of the link above the $i$th mass. Let $\alpha_i$ denote the angle between the line through $(z_{i-1}, y_{i-1})$ and $(z_i, y_i)$ and the $z$-axis; and let $\Delta \alpha_i = \alpha_{i+1} - \alpha_i$.

The force on the $i$th mass, which is in the $y$-direction, is

$$\mathfrak{g} \left( \sum_{j=0}^{i-1} m_j \right) \Delta \alpha_i + \mathfrak{g} m_i \alpha_{i+1}, \tag{9.1}$$

in the small oscillation limit. Euler obtains this force directly by considering horizontal components of the forces of tension in the links and the force of gravity:[4] The $(i-1)$st link exerts the force given by the first term in (9.1)

---

[4] In Euler's notation: for example, the force on the weight $C$ (see fig.) is

$$\frac{C \cdot Cc}{cQ} - \frac{(D+E+F)Dr}{cd}$$

(Cc/cQ equals our $\alpha_4$ and Dr/cd equals our $-\Delta\alpha_3$, in the small oscillation limit, and $C$, $D$, $E$, and $F$ are the weights $m_i$, $i = 3, 2, 1$, and 0.)

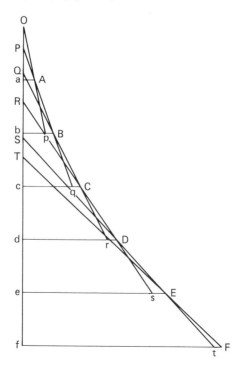

in the horizontal direction while its other component is $g\sum_{j=0}^{i-1} m_j$ in the direction of the $i$th link; gravity exerts the force given by the second term in (9.1) in the horizontal direction while its other component is $g m_i$ in the direction of the $i$th link, in the small oscillation limit. The first term corresponds to the force found by Johann Bernoulli in (8.1) with the weight of the lower masses in the place of tension.

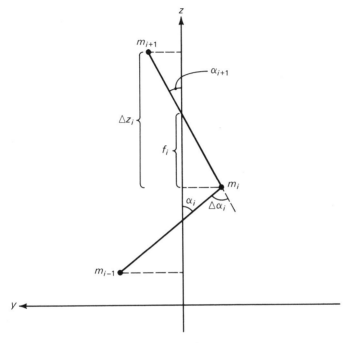

Figure 9.1

Daniel Bernoulli, however, does not follow Taylor and Johann Bernoulli in finding the force from static principles. He deals with the forces through Galileo's law by considering the motions during an infinitesimal period of time. Using induction on the number of masses, he assumes known the motion without the bottom link (so that the bottom particle falls freely) and then he considers the constraining adjustment due to the bottom link. Bernoulli gives his argument in a geometric construction which looks rather complicated because it includes all the infinitesimal separations. We'll give his argument in functional notation:

Suppose that the forces are known for the case of $n$ particles—given by expression (9.1). Bernoulli considers the case of an additional particle of mass $m$ linked to the bottom ($i = 0$) particle at $(0, y_0)$, with link of length $L$ so that its equilibrium position would be $(z_{-1}, 0) = (-L, 0)$. To find the forces in the new system he considers the motion from rest during an infinitesimal time $\delta t$. (This is sufficient since, in the small oscillation limit, the forces are velocity independent.) To find the motion of the new system during time $\delta t$, he first finds its motion without the new link and then he finds the adjustment to this motion caused by forces acting during time $\delta t$ that restore the effect of the new link. First consider the new particle itself: It is to be constrained to the circle $z^2 + y^2 = L^2$ (since the amplitude of the 0th particle is small compared with $L$). If its initial position is $(z, y) = -L\,(\cos \alpha, \sin \alpha)$, then without the effect of the link it moves to $(z - \frac{1}{2}g\delta t^2, y)$. Now the link acts with a force toward the origin sufficient to restore the condition $z^2 + y^2 = L^2$ in time $\delta t$. Whence it is clear that $L\delta\alpha = \frac{1}{2}g \sin \alpha\, \delta t^2$ (see fig. 9.2); so the force along the arc (which can be taken as constant during the time $\delta t$) is $2m(L\delta\alpha/\delta t^2) = mg \sin \alpha$; or, in the case of small oscillations $mg\alpha$, as desired.

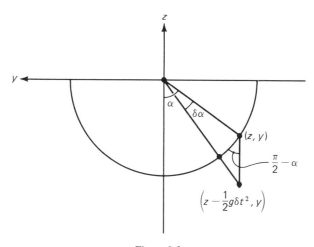

Figure 9.2

Next consider the 0th particle: Without the new link, it moves under the force $m_0 g\alpha_1$ an amount $\delta' y_0 = \frac{1}{2}g\alpha_1 \delta t^2$, along the $y$-axis. The new link, on the other hand, acts on the 0th particle in the direction $-(\cos \alpha, \sin \alpha)$ with magnitude $mg$ (because of the small oscillation limit) and alone would move it by $-mg/2m_0\,(\cos \alpha, \sin \alpha)\delta t^2$. But the original constraints restore it in the direction $(\cos \alpha_1, \sin \alpha_1)$ to the $y$-axis; thus its motion on this axis, due to the new link constrained by the old links would be $\delta'' y = (mg/2m_0)\Delta\alpha\,\delta t^2$, where $\Delta\alpha = \alpha_1 - \alpha$. Therefore the total motion of the 0th particle is

$$\delta y = [\delta'' y + \delta' y] = \tfrac{1}{2}g \left[ \frac{m}{m_0} \Delta\alpha + \alpha_1 \right] \delta t^2$$

and the force is $gm\Delta\alpha + m_0 g\alpha_1$, as desired. Finally, the effect of the new link on the higher particles is found very simply, because of the assumption that the oscillations are small. Thus, the new link causes forces of magnitude $mg$ to act in the directions of the old links. This means that the force $mg\Delta\alpha_i$ is added on to the original force on the $i$th particle, as desired, completing the induction.[5]

Having found the force on the $i$th mass, Bernoulli and Euler impose the pendulum condition:[6]

$$ag\left(\left(\sum_{j=0}^{k-1} m_j\right)\Delta\alpha_k + m_k\alpha_{k+1}\right) = -m_k y_k \qquad (9.2)$$

or

$$ag\left(\left(\sum_{j=0}^{k-1} m_j\right)\left(\frac{y_{k+1}-y_k}{\Delta z_k} - \frac{y_k-y_{k-1}}{\Delta z_{k-1}}\right) + m_k\frac{y_{k+1}-y_k}{\Delta z_k}\right) = -m_k y_k \qquad (9.3)$$

understood with the condition $y_n = 0$. As they indicate, $y_k$ can be expressed in terms of $y_0$ while the condition $y_n = 0$ means that $a$ must satisfy an $n$th degree polynomial. With the $f_i$ defined by $-\alpha_{i+1}f_i = y_i$ (see fig. 9.1), they note that the $k = 0$ equation takes the form[7] $ag = f_0$; and that this means that the bottom mass swings as a simple pendulum about the point $(f_0, 0)$ which is a fixed point where the line of the bottom link crosses the axis. Thus, they find that $f_0$ is the length of the simple isochronous pendulum. Euler also notes that the points $z_i + f_i$, being the intersections of the lines of the links with the $z$-axis, are fixed during the motion. This is clear since the motion only involves scaling of $y$. Bernoulli observes that the oscillating

---

[5] For the case $n = 5$ Bernoulli writes the accelerations of $B$, $C$, $D$, $E$, and $F$ (our $m_i$, $i = 4$, 3, 2, 1, and 0) as

$$p - \frac{C+D+E+F}{B}q$$

$$p+q - \frac{D+E+F}{C}r$$

$$p+q+r - \frac{E+F}{D}s$$

$$p+q+r+s - \frac{F}{E}t$$

$$p+q+r+s+t$$

where $p$ is our $\alpha_5$ and $q$, $r$, $s$, and $t$ are our $-\Delta\alpha_i$, $i = 4, 3, 2$, and 1. Bernoulli's $g = 1$.

[6] In Euler's notation: for $k = 3$, the force written in note 4 is set equal to $C \cdot Cc/f$ where $f$ is our $ag$. But Bernoulli, following the tradition of suppressing constants of proportionality by working with ratios, solves for the *lengths of the isochronous pendula* only when they're needed. Bernoulli solves the cases $n = 2$ and 3 before giving the systematic presentation of the forces (given in note 5).

[7] Euler writes: "The length of the isochronous pendulum $f$ is always equal to the lowest section of the string extended all the way to the vertical 0f, that is, [equal] to the line FT" (see fig. note 4).

linked pendulum has the same shape that it would have were it rotating in equilibrium about its axis. (This observation may have inspired his understanding of the method of the pendulum condition.[8]) For the cases of $n = 2$ or 3, they give detailed discussions and we will come back to this later.

For the case of equal links, $\Delta z_i = \Delta z = l/n$, the pendulum condition (9.3) takes the form[9]

$$\mathfrak{a} g \left( \left( \sum_{j=0}^{k-1} m_j \Delta z \right) \frac{y_{k+1} - 2 y_k + y_{k-1}}{\Delta z^2} + m_k \frac{y_{k+1} - y_k}{\Delta z} \right) = -m_k y_k \qquad (9.4)$$

with the condition $y_n = 0$. They find the equation[10]

$$\mathfrak{a} g \left( \left( \int_0^z \rho(z)\, dz \right) \frac{d^2 y}{dz^2} + \rho(z) \frac{dy}{dz} \right) = -\rho(z) y(z) \qquad (9.5)$$

with the solutions limited by the condition $y(l) = 0$, as the limiting analogue, to describe the shape of a flexible chain (hanging from $(l, 0)$ in the $z$–$y$ plane with linear density $\rho(z)$) undergoing regular oscillations under the pendulum condition.

## 9.2. Laguerre Polynomials and $J_0$

In the discrete case of equal links with equal masses $m_j = \rho l / n$, the pendulum condition (9.4) becomes

$$(k+1) y_{k+1} - \left( 2k + 1 - \frac{l}{\mathfrak{a} g n} \right) y_k + k y_{k-1} = 0 \qquad (9.6)$$

---

[8] In a letter to Euler dated 5 November 1740, Bernoulli wrote: "Your method [for solving the problem of the oscillations of bodies suspended from a flexible thread] is quite close to mine; I have used such a method since the time that I solved the problem of bodies connected by a flexible thread, where I mentioned that when the system is gyrating the shape of the thread must be the same as the shape [it has] in oscillating" (Fuss [1], II, p. 464).

[9] Neither Euler nor Bernoulli actually writes equation (9.4), but they both approach the general continuous case by describing intuitively the general equal link case: Euler writes: "Whether the chain is everywhere of the same thickness or not, the elements 0A, AB, BC, etc. of the curve can, none the less, be taken equal, provided that the weights are chosen correctly for the nature of the chain." Bernoulli writes that "the corpuscles are now to be considered as infinite [in number] and as separated by equal minimal distances, but unequal in weight."

[10] Euler presents this through a term by term translation of the equation for $m_3$ in the discrete case (notes 4, 6) into the corresponding differential elements. In his notation this is $yp\, dt/f = p\, dt\, dy/dt - d\, dy \int p\, dx/dt$, where $p$ is our $\rho$, and $dt$ is our $-dz$; he rewrites this as $yp\, dx^2/f = -d\, dy \int p\, dx - p\, dx\, dy$, where $x$ is our $z$. Bernoulli first writes the equation (e.g. for the uniform case) in the form $\int d\, dy/ds - (l-s) d\, dy/ds^2 = y/n$, where $s$ is our $l - z$, and $n$ is our $\mathfrak{a} g$. He refers to the integral term as "the sum of all the sines of the contact angles [our $d\alpha$]", making the analogy with the discrete case (notes 5 and 9). He then works with the equation in the form $n\, dy\, dx + nx\, d\, dy = -y\, dx^2$, where $x$ is our $z$.

which we recognize as the recursion relation

$$(k+1)L_{k+1}(x)-(2k+1-x)L_k(x)+kL_{k-1}(x)=0 \qquad (9.7)$$

that defines the Laguerre polynomials

$$L_k(x) = \sum_{j=0}^{k} \frac{k(k-1)\cdots(k-j+1)}{j!^2}(-x)^j, \qquad (9.8)$$

(where the coefficient is 1 for $j=0$) uniquely up to a common multiplicative constant. Euler finds this solution to the recursion relation, thus introducing the Laguerre polynomials, solving the problem of the linked pendulum in the case of equal links and equal masses for its shape in a mode:

$$y_k = y_0 L_k\left(\frac{l}{nf_0}\right) \qquad (9.9)$$

where $f_0 = \mathfrak{a}\mathfrak{g}$ can only be chosen so that

$$L_n\left(\frac{l}{nf_0}\right) = 0. \qquad (9.10)$$

To take the limit $n \to \infty$, with $\rho$ and $l$ fixed, first recall that

$$L_k\left(\frac{x}{k}\right) \to J_0(2\sqrt{x}) = \sum_{j=0}^{\infty}\frac{1}{j!^2}(-x)^j \qquad (9.11)$$

as $k \to \infty$ (as entire analytic functions), and the less elementary fact that the $r$th zero of $L_n(x/n)$ converges to the $r$th zero of $J_0(2\sqrt{x})$ as $n \to \infty$. Thus, the respective choices of $f_0 = f_0(n)$ (depending on $n$ through the equality (9.10)) converge as $n \to \infty$ to the respective values of $f_0(\infty)$ with $J_0(2\sqrt{l/f_0(\infty)}) = 0$. Therefore, if $kl/n = z_n \to z$ while $n \to \infty$, it follows from (9.9) and (9.11) that

$$y_k = y_0 L_k\left(\frac{z_n}{kf_0(n)}\right) \to y(z) = y_0 J_0\left(2\sqrt{\frac{z}{f_0(\infty)}}\right). \qquad (9.12)$$

Furthermore $y(z)$ satisfies the pendulum condition (9.5) in the case $\rho(z) = \rho$—that is, with $\mathfrak{a}\mathfrak{g} = f_0(\infty)$ the equation

$$\mathfrak{a}\mathfrak{g}\left(z\frac{d^2y}{dz^2}+\frac{dy}{dz}\right) = -y(z). \qquad (9.13)$$

Now, Bernoulli and Euler are less explicit about the limit; but, as we mentioned above, they obtain the continuous case as the limiting analogue and, as we shall see, they are guided by the fact that the solutions to the condition (9.13) are limits from the discrete case. In particular, the condition (9.13) is their pendulum condition for the small oscillations of a uniform chain. They solve it by the technique, which had become standard, of substituting for $y$ a power expansion in $z$ with undetermined coefficients to obtain a recursion relation for the coefficients which they solve to obtain the series, given via (9.11), for $J_0(2\sqrt{z/f_0})$ with $f_0 = \mathfrak{a}\mathfrak{g}$. But for this to be the shape of a chain oscillating from $(l, 0)$, they note that $f_0$ must satisfy $J_0(2\sqrt{(l/f_0)}) = 0$. We will discuss this further after considering their solutions in the two and three mass cases. Thus they discovered the zeroth order Bessel function and the problem of finding its zeros.

## 9.3. Double and Triple Pendula

Bernoulli and Euler give detailed analyses of the cases of equal links and two and three masses. For $n = 2$, equation (9.4) with the condition $y_2 = 0$, if $\Delta z = b$, becomes

$$
\begin{cases}
y_1 = (1 - b/\mathfrak{a}\mathfrak{g})y_0 \\
\left(2 + (1 - b/\mathfrak{a}\mathfrak{g})\dfrac{m_1}{m_0}\right)y_1 = y_0
\end{cases}
\tag{9.14}
$$

which they solve, obtaining[11]

$$
\begin{cases}
b/\mathfrak{a}\mathfrak{g} = \left(1 + \dfrac{m_0}{m_1}\right) \pm \dfrac{m_0}{m_1}\sqrt{1 + \dfrac{m_1}{m_0}} \\
y_0/y_1 = 1 \pm \sqrt{1 + \dfrac{m_1}{m_0}}.
\end{cases}
\tag{9.15}
$$

For the case $m_1 = m_0$, the displacement ratio is

$$
y_0/y_1 = 1 \pm \sqrt{2}
\tag{9.16}
$$

and $b/\mathfrak{a}\mathfrak{g} = 2 \pm \sqrt{2}$, from which Bernoulli obtains the lengths of the simple isochronous pendula

$$
\mathfrak{a}\mathfrak{g} = \frac{b}{2 \pm \sqrt{2}} \approx \begin{cases} 0.29b \\ 1.71b \end{cases}
\tag{9.17}
$$

and the frequency ratio between the two modes

$$
\sqrt{\frac{2 + \sqrt{2}}{2 - \sqrt{2}}} = 1 + \sqrt{2} \approx 2.41.
\tag{9.18}
$$

Bernoulli writes:

So that I might subject these propositions to experiment, I used two perfectly similar lead balls. ... The balls were connected ... to a steel thread passing through in such a way that the low one was twice as far from the point of suspension as

---

[11] Bernoulli also writes the solutions for the case of unequal links.

was the high one. . . . I drew the lower ball [aside]. . . . Soon the oscillations became uniform and with the help of divisions marked on the wall I ascertained that the displacements of the bodies . . . were as 100 to 241, that is, as 1 to $1+\sqrt{2}$. Also the number of oscillations occurring in a given time corresponded accurately to the length of the simple isochronous pendulum. . . . Next, [the displacement of the lower ball having been made $(1-\sqrt{2})$ times the displacement of the higher ball $I$] released them at the same moment. . . . Again, the number of oscillations agreed accurately with that of the simple isochronous pendulum of length $(1/(2+\sqrt{2}))$ [times the distance from the higher ball to the point of suspension].

We tried this experiment very casually, hanging two cement bricks with string at distances of about a yard from a door frame. Equations (9.17) and (1.3) predict 290 and 120 oscillations in five minutes for the respective modes; and indeed this is what we counted to within 1 or 2 oscillations. That is, it is easy to obtain the ratio given in (9.18) to within a few percent.

Next, for the case of three equal masses with equal links, Bernoulli and Euler have equation (9.6) with $y_3 = 0$, or if we use (9.9), with $q = l/3f_0$,

$$\begin{cases} y_1 = y_0(1-q) \\ y_2 = y_0(1-2q+\tfrac{1}{2}q^2) \\ 0 = 1-3q+\tfrac{3}{2}q^2-\tfrac{1}{6}q^3. \end{cases} \tag{9.19}$$

Thus, they have to solve the third degree Laguerre polynomial for its roots. Actually, they obtain the following polynomial in $x = y_1/y_2$:

$$4x^3 - 12x^2 + 3x + 8 = 0 \tag{9.20}$$

(which follows more directly from (9.6) than from (9.19)) to give the relative positions of the top two masses in the three different modes. Bernoulli also works with the polynomial in $y_0/y_2$. At this point we will pause for a brief interlude to discuss Bernoulli's method for calculating roots.

## 9.4. Roots of Polynomials

A few years before 1728, Bernoulli independently proved[12] de Moivre's theorem that any sequence $s_k = \sum_{i=0}^{n-1} A_i k^i$, defined for given $A_0$, $A_1, \cdots A_{n-1}$, satisfies the recursion relation $s_{k+1} = \sum_{j=0}^{n-1} (-1)^j \binom{n}{j+1} s_{k-j}$. He generalized the converse relation, showing that any sequence $s_k$ that satisfies a recursion relation

$$s_{k+1} = \sum_{j=0}^{n-1} a_{j+1} s_{k-j} \tag{9.21}$$

---

[12] This account of the evolution of ideas is Bernoulli's own, as given in his paper of 1728 (Daniel Bernoulli [2]).

(i.e. a homogeneous, constant coefficient, linear difference equation) can be expressed in the form

$$s_k = \sum_{j=0}^{t} \sum_{i=0}^{m_j-1} A_{ji} \lambda_j^k k^i \tag{9.22}$$

where $\lambda_0, \ldots \lambda_t$ are the roots of the polynomial $\lambda^n - \sum_{j=1}^{n} a_j \lambda^{n-j}$ with respective multiplicities $m_0, m_1, \ldots m_t$ and where the $n$ coefficients $A_{ji}$ can be found in terms of the elements $s_1, s_2, \ldots s_n$. In 1728 Bernoulli observed that if one of the roots has largest modulus, $|\lambda_0| > |\lambda_i|$, $i = 1, 2, \ldots t$, then for typical choices of $s_1, \ldots s_n$, it will dominate in $s_k$, when $k$ is large. Whence he found the convergence $s_{k+1}/s_k \to \lambda_0$, as $k \to \infty$. Replacing $\lambda$ by $1/x$ and multiplying the polynomial by $x^n$, Bernoulli rephrased his method to find the smallest root: If $1 - \sum_{j=1}^{n} a_j x^j$ has a root $r$ of smallest modulus, then it can be found as the limit

$$s_k/s_{k+1} \to r \tag{9.23}$$

where the $s_k$'s are obtained recursively by the relation (9.21). On 20 February 1728, he wrote to Goldbach:[13]

The merit of my method consists in that 1) it is more expeditious, 2) it does nothing by fumbling and 3) that it always shows the same kind of roots, whether the smallest or the largest, as one wishes, ... But even if there were no use for it, I would not be less satisfied, and would consider it as one of the most beautiful theorems that one has in these matters.

Bernoulli used this method to calculate the zeros of the equation (9.20). To give some idea of the amount of work involved, we will apply the method to find the smallest zero: First note that the smallest zero is a negative fraction (one can see this by making various numerical substitutions or else one could know this from a physical sense of a linked pendulum of three masses). Now the equation needs to be considered in the form $1 = -\frac{3}{8}x + \frac{3}{2}x^2 - \frac{1}{2}x^3$ so that $a_1 = -\frac{3}{8}$, $a_2 = \frac{3}{2}$, and $a_3 = -\frac{1}{2}$ in the recursion relation (9.21). Starting with $s_1 = 1$, $s_2 = -1$, $s_3 = 1$, since their ratios give a very rough approximation to the root, one continues with $s_4 = -\frac{3}{8} - \frac{3}{2} - \frac{1}{2} = -2.37500$, $s_5 = \frac{3}{8}s_4 + \frac{3}{2} + \frac{1}{2} = 2.89062$, $s_6 = -5.14648$, $s_7 = 7.45337$, $s_8 = -11.9601$, $s_9 = 18.238$, $s_{10} = -28.506$, $s_{11} = 44.027$, $s_{12} = -68.389$, $s_{13} = 105.9$, $s_{14} = -164.3$, where the ratios, starting with $s_9/s_{10}$ are $-0.640$, $-0.647$, $-0.644$, $-0.646$, $-0.645$. Thus, it is fairly clear that the smallest root, to three significant figures, is $-0.645$. Bernoulli gives this result, while Euler, whose method we don't know, takes less trouble and gives the result as $-0.643$. For the other two roots, Bernoulli obtains[14] 1.353 and 2.292,

---

[13] Fuss [1], II, p. 251.

[14] To find a middle root, Bernoulli preferred the method of dividing the polynomial by the linear factor $(x - r)$, with $r$ approximately the smallest root, to obtain a series whose truncation is a polynomial with smallest root approximately equal to the middle root of the original polynomial.

while Euler agrees on the first but gives the less good result 2.295 on the second. Only Bernoulli calculates the roots of the polynomial that is satisfied by $y_0/y_2$ obtaining the corresponding values 0.122, $-1.044$, and 3.922. Bernoulli does not work directly with the polynomial satisfied by $l/3f_0$ given in (9.19) but he obtains the reciprocals of the roots indirectly from the values of $y_1/y_2$, obtaining respectively $3f_0/l = 2.406$, 0.436, and 0.159. (Accurate values are 2.40515, 0.43587, and 0.158984). He asserts that the corresponding frequencies (see (1.3)) agree well with experiment.[15] Bernoulli seems to have been frustrated in trying to measure the ratios $y_0/y_2$ and $y_1/y_2$ experimentally.

## 9.5. Zeros of $J_0$

We return now to the case of the hanging flexible chain of uniform linear density undergoing small *regular* oscillations. As we saw, Bernoulli and Euler find the curve of the chain to be given by $J_0(2\sqrt{z/f_0})$, with $J_0(2\sqrt{l/f_0}) = 0$. They understand this problem to be analogous to the problem in the discrete case. Bernoulli had generalized his method for approximating the smallest zero of a polynomial to the case of infinite power expansions.[16] The method is this: Consider the power series in the form $1 - \sum_{j=1}^{\infty} a_j x^j$; replace the recursion relation (9.21) by

$$s_{k+1} = \sum_{i=0}^{k-1} a_{i+1} s_{k-i} \qquad (9.24)$$

and use this to define the sequence $s_k$, starting with any initial $s_1, s_2, \ldots s_n$. Provided that the power expansion is absolutely convergent in the neighborhood of the origin and in that neighborhood has a zero of smallest modulus, then for typical choices of $s_1, s_2, \ldots s_n$, the ratios $s_k/s_{k+1}$ will converge to the root. Applying this method to the expansion of $J_0(2\sqrt{l/f_0})$, Bernoulli finds for the reciprocal of the first root $f_0/l = 0.691$. (A good value is 0.69166.) Bernoulli asserts that the frequency $(1/2\pi)\sqrt{g/f_0}$ agrees well with experiment and also that the displacement of the mid-point relative to the bottom, namely $J_0(2\sqrt{l/2f_0})$ agrees with experiment. For his experiment he used a linked pendulum with many small masses and short links. Bernoulli also calculates the reciprocal of the second root obtaining the value $f_0/l = 0.13$. (A good value is 0.131271.) He doesn't mention experimental confirmation. Euler, though he will come back to this problem fifty years later to obtain good values for the first three roots,[17] at this time

---

[15] We tried this casually with bricks and string and got good values for the first two modes but, probably because of the extent of the bricks, we were off by about 7% on the highest mode.

[16] Daniel Bernoulli [3].

[17] Watson [1], pp. 5, 500–501.

makes a calculation of the first root only, obtaining the value $l/f_0 = 1.44$. (A good value is 1.4457965).

Bernoulli and Euler assert that $J_0$ has infinitely many roots; that is, that there are infinitely many decreasing $f_0$ which satisfy $J_0(2\sqrt{l/f_0}) = 0$. Bernoulli goes to some trouble to explain that the corresponding frequencies increase without bound and that the curve of the chain corresponding to the $n$th root has $n - 1$ nodal points (but without mentioning the word "node"), etc. The basis for their assertion of infinitely many roots is, no doubt, based largely on their confidence that the linked pendulum of $n$ bodies has $n$ modes together with their confidence that the continuous case can be understood as a limit. Bernoulli asserts that the distance between the zeros increases with height. The basis for this assertion lies in another remark of Bernoulli's namely that the portion of the chain connecting two adjacent nodal points high up on a long chain (so that its own weight has a negligible effect on its motion in comparison with the tension due to the weight of the lower part of the chain) vibrates as a taut string (in its fundamental mode). From this it follows simply from Mersenne's law that the nodal points must get further apart since the tension increases while the frequency stays the same as one moves up the chain to higher pairs of nodes. Euler makes a remark about an "elegant" geometrical property of the solution: From the equation, $ag(zy'' + y') = -y$, it follows that $\int_0^z y$ is proportional to $-zy'$ which means, if $z$ lies below the first node, that the area between the curve and the axis, below $z$, is proportional to the portion of the $y$-axis between $y(z)$ and the point where the tangent at $(z, y)$ cuts the $y$-axis.

## 9.6. Other Boundary Conditions

Bernoulli takes up the case of the regular oscillations of a uniform flexible chain of length $l$ hung from the bottom of a massless flexible string of length $\lambda$. If $y(z)$ denotes the curve of the entire system of string plus chain, $0 \leq z \leq l + \lambda$, then, because of the flexibility, $dy/dz$ is continuous, even at $z = l$, and $dy/dz$ is constant for $z \geq l$ because of the masslessness of the string. Therefore Bernoulli has the boundary condition[18] $y(l) = -\lambda \, dy/dz]_{z=l}$. The pendulum condition, being concerned only with the forces on the chain is, as before, given by equation (9.13). Thus, Bernoulli finds the curve of the chain, for $0 \leq z \leq l$, to be given by $J_0(2\sqrt{z/f_0})$ as before, but now satisfying the boundary condition

$$J_0(2\sqrt{l/f_0}) + \lambda \frac{d}{dz} J_0(2\sqrt{z/f_0})]_{z=l} = 0 \qquad (9.25)$$

---

[18] In his first paper, theorem 10, Bernoulli states the problem and gives the final series solution; in his second paper he writes only that the derivation follows easily from the methods he has indicated.

which he writes in terms of the power expansion

$$\sum_{j=0}^{\infty} \frac{l^j + \lambda j l^{j-1}}{j!^2} \left( -\frac{1}{f_0} \right)^j = 0. \tag{9.26}$$

Bernoulli calculates the largest reciprocal root for this in the case $\lambda = l$, but he doesn't mention experimental confirmation. The boundary condition (9.25)–(9.26) goes a little against experience with real chains or chords in that one would need two equally flexible chords, the one having negligible linear density compared with the other. Perhaps it was this that motivated Bernoulli to consider the case that we come to next:

Bernoulli modifies the above example by adding a discrete mass $M$ at the point where the string and chain join. Let $Y$ be the displacement of the mass $M$ from the $z$-axis and as usual let $y(z)$, $0 \leqslant z \leqslant l$, be the curve of the chain. Bernoulli states that his previously given methods make it easy to treat also this problem. Thus, we find for the force on $M$, $-(gM/\lambda)Y - gl\rho(-(1/\lambda)Y - dy/dz]_{z=l})$, just as in the case of the top mass in the purely discrete linked pendulum. The force on the element $\rho \, dz$ of the chain is exactly as before, namely $g(\rho z \, d^2y/dz^2 + \rho \, dy/dz) \, dz$. The boundary condition is, of course,

$$y(l) = Y. \tag{9.27}$$

The pendulum condition is standard and becomes

$$\begin{cases} -ag\left(\frac{M}{\lambda}Y + l\rho\left(\frac{1}{\lambda}Y + dy/dz\right]_{z=l}\right)\right) = -MY \\ ag\left(\rho z \frac{d^2y}{dz^2} + \rho \frac{dy}{dz}\right) = -\rho y(z). \end{cases} \tag{9.28}$$

Thus, if $y(0) = 1$, $y(z) = J_0(2\sqrt{z/ag})$, from the second of equations (9.28). Elimination of $Y$ between the first of (9.28) and (9.27) yields

$$-\left(\frac{M+l\rho}{\lambda}\right) J_0(2\sqrt{l/ag}) + \frac{M}{ag} J_0(2\sqrt{l/ag}) - l\rho \frac{d}{dz} J_0(2\sqrt{z/ag})\bigg]_{z=l} = 0. \tag{9.29}$$

As always, Bernoulli works with the power expansions. He writes this condition with the powers of $1/ag$ collected:[19]

$$\sum_{j=0}^{\infty} \left[ \left(\frac{M+l\rho}{\lambda j!^2}\right) l^j + \frac{M}{(j-1)!^2} l^{j-1} + \frac{j\rho}{j!^2} l^j \right] (-1/ag)^j = 0 \tag{9.30}$$

---

[19] In his first paper, theorem 11, Bernoulli expresses our $M$ in the form $\rho L$ so that the $\rho$'s cancel out of the equation. To obtain the series in Bernoulli's actual form with the leading term equal to 1, multiply (9.30) by $\lambda/(L+l)$. There are two misprints in Bernoulli's paper: the third term should have an extra "$+Ll^2$" in the numerator, and in the fifth term, "$16l^3 + L\lambda$" should be "$16l^3 L\lambda$".

(where the term with $(j-1)$ drops out when $j = 0$). In the case $\lambda = l$, $M = l\rho$ he calculates the largest reciprocal root; but he doesn't mention experimental confirmation.

## 9.7. The Bessel Functions $J_\nu$

Finally, Bernoulli treats a non-uniform example of the general continuous case given in equation (9.5) above, namely the case $\rho(z) = \gamma z$. Euler considers the general case $\rho(z) = \gamma z^\nu$. In this case, equation (9.5), the pendulum condition, becomes

$$\frac{z}{\nu+1}\frac{d^2 y}{dz^2} + \frac{dy}{dz} + \frac{1}{\mathfrak{a}\mathfrak{g}} y(z) = 0 \tag{9.31}$$

restricted, as before, by the boundary condition $y(l) = 0$, as well as the implicit condition that $dy/dz$ be bounded. Again by the method of determining coefficients in a power expansion, Euler and Bernoulli solve this equation to find the solution

$$y_\nu(z/\mathfrak{a}\mathfrak{g}) = \sum_{j=0}^{\infty} \frac{1}{j!(\nu+1)\cdots(\nu+j)}(-(\nu+1)z/\mathfrak{a}\mathfrak{g})^j \tag{9.32}$$

with the condition $y_\nu(l/\mathfrak{a}\mathfrak{g}) = 0$. Note that

$$y_\nu(z/\mathfrak{a}\mathfrak{g}) = \Gamma(\nu+1)((\nu+1)z/\mathfrak{a}\mathfrak{g})^{-\nu/2} J_\nu\left(2\sqrt{\frac{(\nu+1)z}{\mathfrak{a}\mathfrak{g}}}\right)$$

$$= \Gamma(\nu+1)\sum_{j=0}^{\infty}\frac{1}{j!\Gamma(j+\nu+1)}(-(\nu+1)z/\mathfrak{a}\mathfrak{g})^j. \tag{9.33}$$

Bernoulli is content in writing out the solution in the $\nu = 1$ and 2 cases, though there is no doubt that he indicates the general method; he calculates the first zero; but he seems to lack the physical motivation for further details. Euler, on the other hand, initiates the study of these functions:

Euler notes that the equation (9.5) takes the form that we call the generalized Riccati equation if one sets $y = \exp\int_0^z (u(z)(\int_0^z \rho(z)\,dz)^{-1})\,dz$, namely

$$du/dz = -\rho(z)/\mathfrak{a}\mathfrak{g} - \left(\int_0^z \rho\,dz\right)^{-1} u^2 \tag{9.34}$$

and in particular that equation (9.31) takes the form

$$du/dz = -\frac{z^\nu}{\mathfrak{a}\mathfrak{g}} - \frac{\nu+1}{z^{\nu+1}} u^2 \tag{9.35}$$

which, he says, is the form proposed by Count Riccati in the case that $\nu$ is half an odd integer. This motivates special attention to the case $\nu = -\frac{1}{2}$ when equation (9.31) takes the form $2zy'' + y' + (\mathfrak{ag})^{-1}y = 0$. This Euler integrates by the standard methods of geometrical calculus. The first integral is $2\mathfrak{ag}zy'^2 = c^2 - y^2$ or

$$\frac{dz}{\sqrt{2\mathfrak{ag}z}} = \pm \frac{dy}{\sqrt{c^2 - y^2}}. \tag{9.36}$$

The integral of the left side is $(2z/\mathfrak{ag})^{1/2}$; the integral of the right side is, if we use functional notation, $+\arccos(y/c) + \text{const.}$, which was geometrically familiar. Not using functional notation, Euler goes to some effort to explain what we now capture by writing

$$y(z) = c \cos((2z/\mathfrak{ag})^{1/2}). \tag{9.37}$$

But he does not give an explanation for the choice of phase, namely, why the constant in the integration is zero. Were it not, one would not have $y'(z)$ bounded at $z = 0$. Euler points out that the zeros are at[20] $z = (\mathfrak{ag}\pi^2/2)(n + \frac{1}{2})^2$, $n = 0, 1, 2 \ldots$; but he leaves it to the reader to choose $\mathfrak{a}$ so that $l$ is a zero. We know from the work of Johann Bernoulli and Taylor, for example, that it was not obvious from the point of view of *geometric* calculus that equation (9.36) (or the simpler equation $dz = dy/(c^2 - y^2)^{1/2}$) has for solution a function $y$ of $z$ that has many zeros. In a sense, this observation is a corollary of the results of Daniel Bernoulli and Euler presently under consideration. That is, since geometric methods have to be supplemented by power expansion techniques, the functional solutions can no longer be overlooked. Thus, it is in the oscillating chain problem and not in the geometrically tractable vibrating string problem that higher modes are discovered mathematically.

Finally in his paper, Euler obtains an integral representation of the function $y_\nu(z)$ defined by the power expansion (9.32). He looks for a solution in the form

$$y_\nu(z) = \frac{1}{H} \int_0^1 t^\alpha (1-t)^\beta R_\gamma(\sqrt{\delta(1-t)z})\, dt \tag{9.38}$$

where

$$R_\gamma(s) = \sum_{j=0}^\infty \frac{\Gamma(\gamma)}{\Gamma(\gamma + 2j)} (-s^2)^j \tag{9.39}$$

with $\alpha, \beta > -1$. For $R_\gamma(s)$ itself, he offers the representation

$$R_\gamma(is) = 1 + e^s s^{1-\gamma} \int_0^s e^{-2r} \int_0^r e^q q^{\gamma-1}\, dq\, dr; \tag{9.40}$$

[20] In giving this result, Euler uses the symbol "$\pi$", defined as $\pi$; but because of an error it should stand for $2\pi$.

though in the end he is only concerned with the case $\gamma = 1$, $R_\gamma(s) = \cos s$. The constant $H$ is chosen by Euler so that $y_\nu(0) = 1$; that is, $H$ is the beta function

$$H(\alpha + 1, \beta + 1) = \int_0^1 t^\alpha (1-t)^\beta \, dt = \frac{\Gamma(\alpha + 1)\Gamma(\beta + 1)}{\Gamma(\alpha + \beta + 2)}. \qquad (9.41)$$

Now, Euler will publish the relationship (9.41) between the beta and gamma functions forty years later;[21] but he already knows the corresponding recursion relations and we can follow him much more conveniently if we use the notation of the gamma function. Euler substitutes the expansion (9.39) into (9.38) and changes the order of summation and integration to obtain

$$y_\nu(z) = 1 + \frac{1}{H} \sum_{j=1}^\infty \frac{\Gamma(\gamma)}{\Gamma(\gamma + 2j)} \left( \int_0^1 t^\alpha (1-t)^{\beta+j} \, dt \right) \frac{\delta^j}{(\nu+1)^j} \left( -(\nu+1)z \right)^j$$

$$= 1 + \sum_{j=1}^\infty \frac{\Gamma(\alpha + \beta + 2)}{\Gamma(\alpha + 1)\Gamma(\beta + 1)} \frac{\Gamma(\gamma)}{\Gamma(\gamma + 2j)}$$

$$\times \frac{\Gamma(\alpha + 1)\Gamma(\beta + 1 + j)}{\Gamma(\alpha + \beta + 2 + j)} \frac{\delta^j}{(\nu+1)^j} \left( -(\nu+1)z \right)^j \qquad (9.42)$$

as we put it, in gamma function notation. Equating terms in the series (9.33) with those in (9.42), we obtain the equality

$$\frac{\delta^j}{(\nu+1)^j} j! \frac{\Gamma(\nu + 1 + j)}{\Gamma(\nu + 1)} \frac{\Gamma(\beta + 1 + j)}{\Gamma(\beta + 1)} = \frac{\Gamma(\alpha + \beta + 2 + j)}{\Gamma(\alpha + \beta + 2)} \frac{\Gamma(\gamma + 2j)}{\Gamma(\gamma)}. \qquad (9.43)$$

This, divided by the corresponding terms with $j - 1$ replacing $j$, yields Euler's equation for the recursion coefficients

$$\frac{\delta}{\nu + 1} j(\nu + j)(\beta + j) = (\alpha + \beta + 1 + j)(\gamma + 2j - 2)(\gamma + 2j - 1). \qquad (9.44)$$

Since (9.44) holds for all $j = 1, 2, 3, \ldots$, being the same as (9.43) in the case $j = 1$, it is equivalent to (9.43). Furthermore the coefficients of the powers of $j$ must be equal. Thus, Euler has the four equations:

$$\begin{cases} \dfrac{\delta}{\nu + 1} = 4 \\[2mm] \dfrac{\delta}{\nu + 1} (\beta + \nu) = 4(\alpha + \beta + 1) + 4\gamma - 6 \\[2mm] \dfrac{\delta}{\nu + 1} \beta\nu = (\alpha + \beta + 1)(4\gamma - 6) + (\gamma - 1)(\gamma - 2) \\[2mm] 0 = (\alpha + \beta + 1)(\gamma - 1)(\gamma - 2) \end{cases} \qquad (9.45)$$

---

[21] Cf. Kline [1], p. 424.

which he solves in the three cases $\gamma = 1$, $\gamma = 2$, and $(\alpha + \beta + 1) = 0$:

CASE I: $\alpha = \nu - \frac{1}{2}$, $\beta = -\frac{1}{2}$, $\gamma = 1$, $\delta/(\nu + 1) = 4$. In this case, (9.38) (with the fact that $R_1(s) = \cos s$) becomes

$$y_\nu(z) = \frac{1}{H} \int_0^1 t^{\nu - 1/2} (1 - t)^{-1/2} \cos \left( 2\sqrt{(\nu + 1)z(1 - t)} \right) dt \qquad (9.46)$$

which Euler writes except that he uses $\sqrt{t}$ for the variable of integration and expresses the cosine in terms of exponentials. Euler assumes that (9.46) is valid under the stated condition $\alpha = \nu - \frac{1}{2} > -1$, or $\nu > -\frac{1}{2}$. (One needs to justify the change in the order of summation and integration. This follows via dominated convergence for $\nu \geq \frac{1}{2}$. For $-\frac{1}{2} < \nu < \frac{1}{2}$, one can first use partial integration.) The integral representation (9.46), used to represent $J_\nu$ through the equality (9.33), is generally called the Poisson–Lommel integral representation and it is valid for Re $(\nu) > -\frac{1}{2}$.

CASE II: $\alpha = \nu - \frac{3}{2}$, $\beta = \frac{1}{2}$, $\gamma = 2$, $\delta/(\nu + 1) = 4$. In this case, Euler's stated condition $\alpha = \nu - \frac{3}{2} > -1$ requires that $\nu > \frac{1}{2}$. He drops this case since it is less general than Case I.

CASE III $i$): $\alpha = -\nu - \frac{3}{2}$, $\beta = \nu + \frac{1}{2}$, $\gamma = 2\nu + 2$, $\delta = 4(\nu + 1)$. In this case, the conditions $\alpha, \beta > -1$ give the restriction $-\frac{3}{2} < \nu < -\frac{1}{2}$ as Euler says and, in addition, since $\gamma \neq 0$, there is the restriction $\nu \neq -1$.

CASE III $ii$): $\alpha = -\nu - \frac{1}{2}$, $\beta = \nu - \frac{1}{2}$, $\gamma = 2\nu + 1$, $\delta = 4(\nu + 1)$. In this case, $-\frac{1}{2} < \nu < +\frac{1}{2}$. Euler drops this case as less general than case I. He also drops case III $i$).

# 10. Euler (1735)

Daniel Bernoulli wrote Euler on 18 December 1734 that he was studying the small (transversal) vibrations of a rod with one end fixed in a wall.[1] On 4 May 1735, he wrote that he had found the equation for its shape, namely $\hat{a}\, d^4y/dz^4 = y$, but that its solutions known to him, namely sine and exponential functions, were inappropriate.[2] Euler responded that he also had found the equation but only the series form of the solution.[3] In October of 1735, Euler presented the St. Petersburg Academy with a paper in which he derived the equation, dealt with the boundary conditions and gave the fundamental solution in series form.[4] This paper is of further significance, however, because in it Euler presented a lucid overview of vibration theory as he understood it at the time. He gave the first explicit discussion of the pendulum condition which, in turn, he looked on as a static equilibrium condition. This reduced dynamical (vibration) problems to static problems which were understood with more sophistication. The paper appeared in 1740.

We will discuss Euler's presentation of the pendulum condition and one of his applications of it to a case of a rigid body oscillating in one degree of freedom in section 10.1. In section 10.2 we will discuss his analysis of the vibrating rod and afterwards, in section 10.3, we will make some brief remarks.

## 10.1. Pendulum Condition

Euler writes that

[In considering] problems which pertain to the oscillations of rigid bodies, Geometers have usually been directed to the discovery of the center of oscillations.

---

[1] Eneström [2], p. 144. Bernoulli did not describe his solution, but wrote that he was not satisfied with it.

[2] Fuss [1], II, 422. Bernoulli wrote: "For the curve [of the vibrating elastic lamina] I find the equation $n\, d^4y = y\, dx^4,\ldots$ But this matter is very slippery, and I would like to hear your opinion about it. The logarithm satisfies both the above equation and the equation $n^{1/2}d\, dy = y\, dx^2$, but it is not general enough for the present business. You will already have observed that $n\, d^my = y\, dx^m$ contains in itself as particular cases $\alpha\, d^py = y\, dx^p$, where $p$ is a divisor of $m$."

[3] We have not seen this letter. It is described by Truesdell [1], p. 167.

[4] Euler [5].

[But] a double question is to be raised about the oscillations of flexible bodies. For before it is possible to determine the length of a simple isochronous pendulum, the shape which a flexible body assumes [when vibrating] must be found; for unless this is known, the forces acting cannot be assigned.

He goes on to say that Taylor was the first to solve a problem of this kind and he mentions the hanging chain problem which he and Daniel Bernoulli had recently solved. He states that all previous work on oscillations involved diverse methods which did not help him to solve the vibrating rod problem.

[But] then I thought of a clear method based on static principles, with the help of which I not only solved with wonderful ease the questions of the elastic rod and the hanging chain, but also I was promptly able to clear up all questions pertaining to oscillations.

After a pedagogical discussion of the simple pendulum, Euler describes his method:

In all cases of oscillations the state of equilibrium must be considered first of all. . . . Next the body is perturbed from that state and one is to find by what interval each element [of the body] is moved. . . . Finally, so that all elements return simultaneously to their equilibrium locations, the force on each element must be as its mass times the length of the interval to be travelled [in returning to the equilibrium position]. . . .

This is, of course, the, by now standard though not yet exposited, pendulum condition, motivated through the assumption of isochronism and simultaneous crossing of the axis (except that the "axis" is generalized to the configuration of equilibrium). Euler also puts this differently by saying that the harmonic forces can be "substituted" for the actual forces acting. The second part of Euler's method reduces dynamics to statics:

Since the forces to be substituted are equivalent to the actual forces, it is known from statics that if equal but opposite forces [to those substituted] are instead added to the actual forces, the body must be in equilibrium [in its perturbed configuration]. . . . Therefore . . . everything concerning the oscillations of bodies is reduced to static principles.

Thus, Euler looks on the pendulum condition (1.2) as a condition of static equilibrium:

$$F + \mathfrak{a}^{-1} m y = 0. \tag{10.1}$$

Though we have been avoiding problems that involve only a single degree of freedom, we will pause briefly to follow Euler's application of his method to the problem of rocking. His results on rocking seem to be new, although Daniel and Johann Bernoulli obtained them independently.[5]

---

[5] On 26 October 1735, Daniel Bernoulli wrote Euler that also he had considered the infinitely small "oscillations of a cradle" (Fuss [1], II, 429–430). His result agrees with Euler's. See also Chapter 15, section 1.

Consider the small oscillations of a body with a convex base which rocks on a plane surface without slipping. Suppose that the body belongs to the $x$–$y$ plane so that in equilibrium its mass density is $\rho(x, y)$ and its base is a curve in the upper half-plane tangent to the $x$-axis at the origin where it has center of curvature $(0, R)$. If the body rocks on the $x$-axis through an angle $\alpha$ which is small, the element at $(0, R)$ moves to $(-R\alpha, R)$ and the point of contact with the $x$-axis moves to $(-R\alpha, 0)$. Furthermore, an arbitrary element of the body at $(x, y)$ is displaced by $(\delta x, \delta y) = \alpha(-y, x)$ and hence, by the pendulum condition, will experience the harmonic force[6] $-\mathfrak{a}^{-1}\rho\, dx\, dy\, \alpha(-y, x)$, where $\rho$ is evaluated at $(x, y)$. Therefore the forces $+\mathfrak{a}^{-1}\rho\, dx\, dy\, \alpha(-y, x)$ along with the gravitational forces $\mathfrak{g}\rho\, dx\, dy\, (0, -1)$, the forces of rigidity and the lifting and frictional forces of the $x$-axis hold the body in equilibrium in its rotated state. The total force is clearly zero because the force exerted by the $x$-axis balances by definition. Therefore a necessary and sufficient condition of equilibrium is that the forces have no total moment about $(-R\alpha, 0)$. That is, since the lifting, frictional and rigidity forces don't contribute, that[7]

$$0 = \int\int \{\mathfrak{a}^{-1}\rho\alpha(-y, x) + \mathfrak{g}\rho(0, -1)\} \wedge (x + \delta x + R\alpha, y + \delta y)\, dx\, dy \quad (10.2)$$

or[8]

$$0 = \int\int \{\mathfrak{a}^{-1}\rho(x^2 + y^2) - \mathfrak{g}\rho(x/\alpha - y + R)\}\, dx\, dy \quad (10.3)$$

without second order terms in $\alpha$; or, with $M = \iint \rho\, dx\, dy$, $I_0 = \iint (x^2 + y^2)\rho\, dx\, dy$ and $Y = M^{-1} \iint y\rho\, dx\, dy$, since $\iint x\rho\, dx\, dy = 0$, the length of the simple isochronous pendulum is[9]

$$\mathfrak{a}\mathfrak{g} = I_0/M[R - Y]. \quad (10.4)$$

Euler observes that if the direction of the gravitational field is reversed and the point at which the body is held is fixed at the origin, he has the case of the compound pendulum. This modifies (10.3) and (10.4) by removing $R$ and changing the sign of $\mathfrak{g}$. Hence the length of the simple pendulum isochronous with the compound pendulum is

$$\mathfrak{a}_s\mathfrak{g} = I_0/MY, \quad (10.5)$$

_____

[6] In Euler's notation: the force is $M \cdot Mm/f$, where $M$ is our $\mathfrak{g}\rho\, dx\, dy$, $Mm$ is the length of our $\alpha(-y, x)$, and $f$ is our $\mathfrak{a}\mathfrak{g}$.

[7] Euler's procedure differs only in that he first obtains separately the moments due to the gravitational and the harmonic forces and then equates their magnitudes. (Although Euler uses geometric notation, he obtains the moments in terms of the cartesian components of the forces and distances, that is, in the same terms that appear in the indicated cross product.)

[8] In Euler's notation: the moment due to the gravitational force is $\int M \cdot PA + \int M \cdot CQ \cdot A\alpha/CA$, where PA is our $x$, CQ is our $R - y$, CA is our $R$, and $A\alpha$ is our $R\alpha$; the moment due to the harmonic force is $\int M \cdot AM^2 \cdot A\alpha/CA \cdot f$, where $AM^2$ is our $(x^2 + y^2)$.

[9] In Euler's notation: $f = \int M \cdot AM^2 / \int M \cdot CQ$, where $\int M \cdot AM^2$ is our $I_0$, and $\int M \cdot CQ$ is our $\mathfrak{g}M[R - Y]$.

the well known result of Huygens. This gives Euler the nice geometrical result[10] that

$$a/a_s = Y/[R - Y],\qquad(10.6)$$

which makes the result for rocking memorable.

## 10.2. Vibrating Rod

To begin his discussion of the vibrating rod, Euler recalls the paper that he has written seven years earlier on statics of the rod.[11] Here he considered a one-dimensional rod (that doesn't undergo any stretching or twisting) that is straight in its unstressed state and can be bent in the plane. He considered the case in which it is in equilibrium with one end clamped while an arbitrary force acts at the other and an arbitrary force density acts along its length. Suppose that the unstressed rod lies along the interval $0 \leqslant z \leqslant l$ in the $z$–$y$ plane, clamped at $z = l$; and suppose that in its stressed equilibrium it lies along the curve $(z(s), y(s))$, where $s \in [0, l]$ is arc length. Let the force density be denoted $(q(s), p(s))$ and let the end force be $(F, E)$. Euler found the bending moment at $(z_0, y_0) = (z(s_0), y(s_0))$ as the negative of the moment due to the forces acting to the left of the point:

$$\mathcal{M}(s_0) = -\int_0^{s_0} (z(s) - z_0, y(s) - y_0) \wedge (q(s), p(s))\, ds$$

$$- (z(0) - z_0, y(0) - y_0) \wedge (F, E)$$

$$= -\int_0^{s_0} \{z - z_0\} p\, ds + \int_0^{s_0} \{y - y_0\} q\, ds$$

$$- \{z(0) - z_0\} E + \{y(0) - y_0\} F.\qquad(10.7)$$

Euler supposed that $p$ and $q$ can be considered as functions of $z$ and $y$ respectively; so, by a partial integration which he carried out geometrically, he obtained the bending moment

$$\mathcal{M}(s_0) = \int_{z(0)}^{z_0} \left\{ \int_{z(0)}^{z} p\, dz \right\} dz - \int_{y(0)}^{y_0} \left\{ \int_{y(0)}^{y} q\, dy \right\} dy$$

$$+ \{z_0 - z(0)\} E - \{y_0 - y(0)\} F.\qquad(10.8)$$

On the other hand, Euler asserted that the bending moment is proportional to the curvature

$$\mathcal{M}(s_0) = e\kappa(s_0),\qquad(10.9)$$

---

[10] In Euler's notation: $F/f = CG/AG$, where $F$ is our $a_s$, CG is our $R = Y$, and AG is our $Y$.

[11] Euler [4].

$\kappa$ being the curvature of the curve and $e$ being what we would call the flexural rigidity. A little earlier, in his work on the vibrating ring (section 7.1), Euler had made a detailed local analysis of the internal forces of the rod. But he justified (10.9) only as the limiting analogue of the discrete case in which bending can occur only at joints that are separated by rigid segments. In this case, (10.9) is

$$\mathcal{M}_i = (e/\Delta s_i)\Delta\alpha_i \qquad (10.10)$$

(in the notation that we used in Chapters 8 and 9). This is the assertion, given as an "Hypothesis" at the beginning of the paper, that the moment at a joint is proportional to the angle through which the joint is turned. Euler wrote that it is commonly assumed and that it can probably be argued in terms of physical principles. In any case, Euler had from (10.8) and (10.9) the equilibrium equation with which he works.[12]

Now, in the case of the small vibrations of a rod clamped at one end, Euler can apply his method. In this case, $z = s$ and $\kappa = d^2y/dz^2$ and $q$, $F$ and $E$ are all zero. The pendulum condition is then given by (10.8) and (10.9) with $p(z) = \mathfrak{a}^{-1}\rho(z)y(z)$. That is,[13]

$$\mathfrak{a}e\, d^2y/dz^2 = \int_0^z \left\{ \int_0^z \rho y \, dz \right\} dz. \qquad (10.11)$$

Euler has immediately the conditions $y(l) = 0$ and that $y(0)$ is prescribed as the amplitude; he "finally understands" that the final constant of integration is determined by the condition $dy/dz]_{z=l} = 0$, corresponding to the fact that the rod is clamped. Euler notes, of course, that on two differentiations (10.11) becomes

$$\mathfrak{a}e\, d^4y/dz^4 = \rho y; \qquad (10.12)$$

but he prefers the form (10.11) because it takes care of two constants of integration (disallowing, for example, the sine and exponential functions) and also perhaps because he doesn't see $e\, d^4y/dz^4$ as a natural force density while (10.11) deals directly with physical moments. But Euler does see that he can work with (10.12) if he requires the solutions to satisfy (10.11) (i.e. two additional conditions $d^2y/dz^2]_{z=0} = 0$ and $d^3y/dz^3]_{z=0} = 0$). This work, along with the works on the hanging chain (Chapter 9) seems to represent in part at least the discovery of boundary conditions.

Euler makes a very limited effort to solve (10.11) or (10.12) subject to its boundary conditions in the case of a constant mass density, $\rho = 1$; he indicates only the most standard series manipulations. Thus, if $y = \sum c_i z^i$,

---

[12] In Euler's notation: $V/r = Ex + Fy + \int P\, dx + \int Q\, dy$, where $V$ is the "elastic force" (our $e$), $1/r$ is our $\kappa$, $x$ is our $z_0 - z(0)$, $y$ is our $-y_0 + y(0)$, $P$ is our $\int p\, dz$, and $Q$ is our $\int q\, dy$.

[13] In Euler's notation: $A\, du = dx^2 \int dx \int u\, dx/f$, where $A$ is our $e/\rho$, $u$ is our $y$, and $x$ is our $z$.

then (10.12) gives $ae(j+1)(j+2)(j+3)(j+4)c_{j+4} = c_j$; so the conditions $d^2y/dz^2]_{z=0} = 0 = d^3y/dz^3]_{z=0}$ limit the series to the form

$$y = c_0 + c_1 z + \frac{c_0}{ae\,4!}z^4 + \frac{c_1}{ae\,5!}z^5 + \frac{c_0}{(ae)^2 8!}z^8 + \frac{c_1}{(ae)^2 9!}z^9 + \cdots. \quad (10.13)$$

The constant $c_0$ is given by the amplitude $y(0)$. Euler then indicates that $c_1$ is to be eliminated through the substitutions $y(l) = 0$ and $dy/dz]_{z=l} = 0$ in (10.13); $c_0$ then cancels out and one obtains an expression which is to be solved for the harmonic constant $a$. He indicates only the fundamental solution which he gives approximately[14] and he makes no mention of the possibility of higher modes.

Euler is most interested in the universality of this method. Thus he gives a presentation of the vibrating string from the same point of view. In this case, the pendulum condition is again given by (10.8) and (10.9) with $p(z) = a^{-1}\rho y(z)$, and $z = s$, but now with $e = 0$, $q = 0$, and $F = -P$ the tension. This gives the equation

$$0 = \int_0^z \left\{ \int_0^z \rho y \, dz \right\} dz + aPy + aEz. \quad (10.14)$$

The force component $E = -Py'(0)$, which one sees by differentiating (10.14) once. Euler differentiates (10.14) twice to obtain the standard equation for the vibrating string which he solves for the fundamental mode only, pointing out his agreement with Taylor and Johann Bernoulli but asserting that he used "quite different principles."

Euler concludes his paper saying that equation (10.11) is appropriate for a rod that is hinged at both ends but that it must be "accommodated in another way in this case."

## 10.3. Remarks

It may be worth remarking on a few things that Euler did not do in this paper: 1) He didn't notice that $[-d^2\mathcal{M}/dz^2]\,dz$ is the force on the element $dz$ of the rod due to the stress of bending. This follows for example from (10.8). Had he done so he could have obtained the differential equation (10.12) directly from (10.9) and the pendulum condition, that is, in exactly the same way that he obtained the equation for the hanging chain. But he writes

However much I obtained an easy and general method for finding the curve of a perfectly flexible oscillating chain, nevertheless, from that method I was able to

---

[14] Bernoulli's treatment, which is more detailed, will be discussed in Chapter 13.

obtain little use for finding the curve of the oscillating elastic lamina. Then I thought of a clear method based on static principles. . . .

2) He didn't notice that the solutions of (10.12) can be written in the form

$$A \sin cz + B \cos cz + C \exp cz + D \exp -cz$$

where $c^4 = ae/\rho$ and where the constants $a$, $A$, $B$, $C$, and $D$ can be determined from the boundary conditions and the amplitude (allowing for different modes). 3) He didn't notice or, at any rate, mention the possibility of higher mode solutions. This seems a little strange since he mentions having already done his work on the hanging chain (though it was published in a later volume) and in any case he had seen Daniel Bernoulli's work on the hanging chain.

In the appendix to his book of 1744 on the calculus of variations,[15] Euler gives a systematic and readable presentation of the vibrating rod. For each set of boundary conditions he finds all the modes in the form indicated in the remark 2) above (except that he excludes all the odd modes in the case of the rod free at both ends when he should exclude only the first). His starting point—the pendulum condition as a static equilibrium problem—remains the same. Here he explicitly poses the problem of the vibrating rod as a problem to be solved by the methods of the calculus of variations but he doesn't know what should be minimized to do so.

---

[15] Euler [9].

# 11.  Johann II Bernoulli (1736)

Johann II Bernoulli's essay on light propagation, which received a prize
of the Paris *Académie*, was published[1] in 1736. In this essay Bernoulli
attempts to describe light as a pressure wave in an elastic medium that can
be conceived as a sprinkling of small particles in an ether full of very small
vortices.[2] A portion of the essay is devoted to an analysis of the propagating
pressure wave. As had Newton, he says that the analysis is applicable to
sound as well as to light and he attempts to derive the velocity of sound.
He obtains the forces in a slightly deformed elastic medium, following
Newton except that he begins by considering a discrete version of the
medium. Thus he finds for the forces an expression which is the same as
that found by Johann I Bernoulli in the case of the loaded string. In order
to complete the analogy with the string he associates the half wave length
with string length; but while his model would be appropriate for standing
waves, he misapplies it to describe propagating waves. He argues that a
modified version of the model would account for the experimental discrepancy in Newton's result for the speed of sound. We will present his
discussion of the pressure wave and its modified version; in section 11.1
we will make some brief remarks.

## 11.1.  Pressure Wave

Bernoulli considers the propagation of a pressure wave along a ray that
emanates from the source; but effectively he considers the ray as a one-
dimensional gas. In our usual notation, let $z$ label a point of this gas with
the source of sound at $z = 0$. Bernoulli begins with a discretized version
of the gas so that, in equilibrium, equal masses $\rho \, \Delta z$ lie at the points $z_i$
which are spaced at equal intervals $\Delta z = z_{i+1} - z_i$ and between the masses
there is a massless gas that satisfies Boyle's law. Thus, if $y_i$ is the deviation

---

[1] Johann II Bernoulli [1]; the relevant discussion of the propagation is in pars. XLIX–
LXXV, "Explication analytique de la nature et du mouvement des fibres lumineuses et des
fibres sonores."

[2] Bernoulli's model of the ether is described by Whittaker [1], pp. 95–96.

of the $j$th mass from its equilibrium position $z_j$, he supposes the $(i-1)$st and $i$th masses to repel each other with a force

$$\left| P\frac{\Delta z}{\Delta z + y_i - y_{i-1}} \right|.$$

He has then that the force on the $i$th particle is[3]

$$P\,\Delta z\left(\frac{1}{\Delta z + y_i - y_{i-1}} - \frac{1}{\Delta z + y_{i+1} - y_i}\right) = P\,\Delta z\,\frac{y_{i+1} - 2y_i + y_{i-1}}{(\Delta z)^2} \quad (11.1)$$

in the small oscillation limit, $|y_i| \ll \Delta z$. This is exactly the same as the force on the $i$th mass of a loaded string found by Johann I Bernoulli provided the pressure $P$ is replaced by the tension $P$.

Bernoulli believed that the motion of small displacements in a gas must be simple harmonic (see section 11.2). Therefore and for no other reason, on obtaining the force in the discrete case, he gives immediately the pendulum condition[4]

$$\hat{a}\,\frac{y_{i+1} - 2y_i + y_{i-1}}{(\Delta z)^2} = -y_i, \qquad \hat{a} = \mathfrak{a}P/\rho. \quad (11.2)$$

He doesn't discuss the force in the continuous case but goes directly to the corresponding pendulum condition[5]

$$\hat{a}\,\frac{d^2 y}{dz^2} = -y, \qquad \hat{a} = \mathfrak{a}P/\rho. \quad (11.3)$$

Thus, following his father, Bernoulli has implicitly that the force on $\rho\,dz$, in the continuous case, is $P(d^2y/dz^2)\,dz$, which is the force found by Newton in (2.1) and (2.2).

Bernoulli attempts to explain the analogy between the pressure wave and the vibrating string. To do this, he views a propagating pressure wave, somewhat as Hermann had, to be like a standing wave whose agitation propagates. In particular, this view provides him with nodal points analogous to the fixed end points of the vibrating string. Bernoulli describes this as follows: If a displacement is made at the origin, a similar displacement (due to the forces described above) will occur on each side of the origin and this displacement will decrease with distance from the origin until finally at $\pm l/2$, say, there will be no displacement at all. Thus he completes the analogy with the vibrating string by setting the string length equal to $l$. He explains propagation by asserting that, while there is no displacement

---

[3] In Bernoulli's notation: the force per mass density is $\mathfrak{g}A(1/(a+r-s) - 1/(a+s-t))$ or, in the small vibration limit, $\mathfrak{g}Aa(2s-t-r)/a^2$, where $\mathfrak{g}A$ is our $P/\rho$, $a$ is our $-\Delta z$, $r$, $s$, & $t$ are our $y_{i+1}$, $y_i$, & $y_{i-1}$.
[4] In Bernoulli's terminology: the intensity of the accelerating force, $\mathfrak{g}A(2s-t-r)/a^2 s$, must be the same for all the particles (namely $\mathfrak{a}^{-1}$ in our notation).
[5] In Bernoulli's notation: $-(d\,dt/t) = (dy^2/c^2)$, where $t$ is our $y$, $y$ is our $z$, and $c^2$ is our $\hat{a}$.

at $\pm l/2$, there is actually enough displacement there to cause a motion in the intervals $(-3l/2, -l/2)$ and $(l/2, 3l/2)$ equal but opposite to the motion in $(-l/2, l/2)$. After another half cycle, he has it that the nodal points $\pm 3l/2$ cause similar motions in $(-5l/2, -3l/2)$ and $(3l/2, 5l/2)$, etc. This description of propagation does lead to the correct expression for velocity of propagation

$$V = 2l\nu \qquad (11.4)$$

where $\nu$ is the frequency. Bernoulli has from his father's and Taylor's analysis of the vibrating string that the frequency is given by (3.6), $\nu = (1/2l)\sqrt{P/\rho}$, and therefore from (11.4) he obtains Newton's result[6] (2.4),

$$V = \sqrt{P/\rho}. \qquad (11.5)$$

Bernoulli calls the gas segments $(-l/2, l/2)$, $(l/2, 3l/2)$, etc. fibres and he elevates their physical status, suggesting for example that a fibre of a certain length involves a resonance of gas particles of a certain size. Seeing the fibres as real in their own right, Bernoulli is able to consider a modification of their pendulum condition that need not be fully justified from the original point of view. His motivation for such modification is the experimental discrepancy in Newton's velocity (11.5). The measured velocity of sound being higher than Newton's calculated velocity, Bernoulli wants the vibrating fibres to have a smaller period. Thus he suggests that a fibre should be described as a "musical string" in the shape of a double cone with common apex in the center and bases at the end points but otherwise unchanged from the uniform string. This presumably means that the tension and total mass are unchanged so that the uniform density $\rho$ is changed to $12\rho l^{-2}z^2$ for $-l/2 \leqslant z \leqslant l/2$. This leads to the new pendulum condition[7]

$$\hat{a}_B \frac{d^2y}{dz^2} + z^2 y = 0, \qquad \hat{a}_B = a_B Pl^2/12\rho \qquad (11.6)$$

($a_B$ being the new harmonic constant). Bernoulli says only that he has made approximations which indicate that $a_B$ is smaller as desired. But it is much too small, as one can see using the techniques of Daniel Bernoulli and Euler (Chapter 9): $y(z)$ is to satisfy (11.6) so that $y'(0) = 0$ and $l/2$ is the first zero, $y(l/2) = 0$. Thus, up to a constant factor,

$$y(z) = K(\hat{a}_B^{-1/4} z) \qquad (11.7)$$

---

[6] In Bernoulli's notation: the velocity of sound is $p\sqrt{DA}$, where $D$ is the length of a second's pendulum whose period is two seconds ($D = g/\pi^2$), $A$ is our $P/g\rho$, and $p$ is $\pi$.

[7] In Bernoulli's notation: $-(d\,dt/t) = (y^2\,dy^2/c^4)$, where $t$ is our $y$, $y$ is our $z$, and $c^4$ is our $\hat{a}_B$. Bernoulli does nothing with this equation; in particular we can't verify explicitly the connection between $\hat{a}_B$ and $a_B$ given in (11.6).

with $K$ satisfying $K'' + z^2 K = 0$, $K'(0) = 0$, $K(0) = 1$, and $\hat{a}_B$ fixed so that $\hat{a}_B^{-1/4} l/2$ is the first zero of $K$. The standard method of undetermined coefficients gives

$$K(z) = 1 - \frac{1}{4 \cdot 3} z^4 + \frac{1}{8 \cdot 7 \cdot 4 \cdot 3} z^8 - \frac{1}{12 \cdot 11 \cdot 8 \cdot 7 \cdot 4 \cdot 3} z^{12} + \cdots . \quad (11.8)$$

The first zero can be approximated by Daniel Bernoulli's method (section 9.5); it is about 2.00. Thus, $\hat{a}_B \approx l^4/256$ and the harmonic constant $a_B \approx (3/64) l^2 P/\rho = (3\pi^2/64)a$, by (3.5). Therefore Bernoulli would obtain a new frequency and hence a new velocity of sound propagation that is greater by a factor of about $8/\pi\sqrt{3} = 1.47$. But Newton's result should be corrected by a factor of about 1.18.

## 11.2. Remarks

Bernoulli motivates his conical fibre by referring to radial propagation from a central source. Of course, had he derived an equation from this point of view he would have obtained the equation for radial eigenfunctions of the Laplacian, which transforms into the original equation and does not yield any change in velocity. Still, Bernoulli's motivation makes some sense for the initial fibre; but since the other fibres which join end to end have to satisfy the same equation, his motivation is strange.

Bernoulli accepts Taylor's explanation of the paradox:

One must not think that, when [a] particle . . . begins to be agitated or vibrated, all the other particles which are to form the fibre acquire at that same moment this conspiring motion necessary for each particle to make its oscillations conjointly with the others; but nevertheless, this first irregularity which would prevent . . . iso-chronous vibrations ends quickly, [the particles] accommodating themselves to each other much in the manner that a tense musical string takes on any figure that one wants to give it, for example, that of an isosceles triangle, when one displaces it from its axis, but gives up this figure after a small number of vibrations, when one lets go of it, converging very promptly to the curvature of the [sine].

Bernoulli emphasizes the approximation in (11.1) and generalizes it to the assertion that the forces will be linear, as we say, in the case of small deviations from "forced" equilibrium. (In the case at hand, the equilibrium is "forced" because the $i$th particle is pushed from both the right and the left.) As Bernoulli says, a linear force leads to isochronous vibrations in the familiar case, namely the case of one degree of freedom. However, he applies this to say, in effect, that linear forces lead to isochronous vibrations in general.[8] Thus, according to him, vibrations in a gas must be isochronous.

---

[8] Bernoulli argues this in a "General proposition" earlier in the essay (pars. XXXVIII–XL).

On the basis of this, he criticizes Newton for assuming, rather than proving, that the motion in a propagating wave is simple harmonic,[9] following Cotes and Hermann in their misunderstanding.

Johann I Bernoulli was convinced that his son's use of the conical string corrected Newton's result for the velocity of sound. On 2 April 1737 he sent Euler a copy of the essay.[10] In his reply of 27 August, Euler was doubtful, preferring his own earlier (incorrect) derivation.[11] Then in a letter of 6 November, Johann I Bernoulli tried to convince Euler of the model's validity.[12]

Johann II Bernoulli may have taken his physical interpretation of the fibre as involving resonances of gas particles of a certain size from Mairan's hypothesis that particles of different elasticities are associated with the propagation of different pitches (section 6.2). At any rate, when Mairan came back to his hypothesis and presented it in great detail to the Paris *Académie* on 4 May 1737, he remarked that his ideas had "been adopted and developed with much elegance" in Johann II Bernoulli's essay. In his essay, Bernoulli writes:[13]

Although the corpuscles swim helter-skelter in the ether, large ones along with small, ... one must imagine that when the agitation arrives, the corpuscles separate and arrange themselves in such a way that each fibre is composed of equal ones. ... [The size of a fibre's first corpuscle] suffices to determine that all the corpuscles of the same size that are found between the two extremities of the fibre remain there and begin to participate in the agitation.... Larger or smaller ones, not being suited to follow the original agitation as easily, will be expelled.... These sorts of communicative movements in bodies having the same disposition to movement are something quite ordinary in nature: we see, for example, that [when] musical strings [are] stretched close to each other, ... one of these, being plucked, will cause all the others that are tuned to the same pitch to tremble....

Johann II Bernoulli's discussion of the analogy between the vibrating string and pressure waves was wrong because he forced it to apply to propagating waves. The analogy would be correct for standing waves. This fact, along with the observation that standing waves occur in musical pipes, was recognized by Daniel Bernoulli in later work that was probably

---

[9] Bernouli writes:
... to confess the truth, [Newton's] long argument in propositions 47, 48, 49 ... and also in the scholium is so obscure and so perplexing to me that I cannot brag that I understand it, especially the way he argues in prop. 47, where it seems difficult to disentangle that which he assumes from that which he wants to prove.

[10] Fuss [1], II, p. 12.

[11] Eneström [1], V, p. 256. Euler was presumably referring to the derivation that had led him to the result presented in his dissertation of 1727 (see Chapter 7, section 2).

[12] *ibid.*, p. 263.

[13] par. XXXIII.

conceived a few years after the appearance of Johann II Bernoulli's discussion.[14] However, even Daniel Bernoulli followed his brother in conceiving of a propagating wave as having nodal points.[15]

---

[14] Daniel Bernoulli wrote to Euler on 7 March 1739 that he had a new theory for the sound of flutes and that it extended also to conical pipes (Fuss [1], II, p. 454). He published the work in Daniel Bernoulli [12].

[15] Daniel Bernoulli [12], par. 47.

# 12. Daniel Bernoulli (1739; 1740)

Daniel Bernoulli presented the St. Petersburg Academy with papers on the bobbing and rocking of a floating body and on the oscillations of a rigid body hanging from a string. These papers appeared in 1750, in the volumes for 1739 and 1740, respectively.[1,2] Bernoulli treats the case of an asymmetric floating body so that its regular motions (simple modes) involve both bobbing and rocking. We will discuss this work in section 12.1 and make some historical remarks on it in section 12.2. In section 12.3, we will discuss Bernoulli's treatment of the dangling rod. The problems of the floating body and the dangling rod are physically interesting and observable while mathematically they are especially tractable, involving only two degrees of freedom. Yet Bernoulli makes no new discoveries about dynamics or vibration theory; he gives elegant applications of what he has already learned in his considerations of the hanging chain. In section 12.4, we will remark on the fact that superposition of simple modes is not discovered.

## 12.1. Floating Body

Bernoulli considers a two dimensional rigid body which is acted on only by the force of gravity and the buoyancy force of a two dimensional incompressible fluid. In equilibrium, let the body occupy the set $\mathcal{B}$ in the $x$–$y$ plane; let its mass density be denoted $\rho(x, y)$; suppose that gravity acts in the direction $(0, -1)$ and that the fluid, having density $D$, lies in the lower half-plane. Let $\mathcal{S}$ denote the submerged part of $\mathcal{B}$ i.e. the intersection of $\mathcal{B}$ with the lower half-plane. The mass of the body is

$$M = \iint_{\mathcal{B}} \rho \, dx \, dy. \qquad (12.1)$$

---

[1] Daniel Bernoulli [7].
[2] Daniel Bernoulli [8].

Let the center of mass lie on the $y$-axis:

$$\mathbf{G} = (0, -G) = \frac{1}{M} \iint_{\mathcal{B}} (x, y)\rho \, dx \, dy \tag{12.2}$$

(where $-G$ could be positive). Since $\mathcal{B}$ is in equilibrium,

$$M = D \iint_{\mathcal{S}} dx \, dy \tag{12.3}$$

and the center of buoyancy lies on the $y$-axis,

$$\mathbf{B} = (0, -B) = \frac{D}{M} \iint_{\mathcal{S}} (x, y) \, dx \, dy. \tag{12.4}$$

If the equilibrium is stable and the body is slightly displaced, then it will rock and bob. Since no lateral forces are involved, the center of mass will remain on the $y$-axis if it is not removed initially. Thus, we can parametrize displacement with the parameters $\alpha$, the rotation about the center of mass, and $\beta$, the displacement along the $y$-axis of the center of mass. To first order in $\alpha$, a point $(x, y)$ of the body is displaced by

$$\alpha(-y - G, x) + \beta(0, 1). \tag{12.5}$$

Let $\mathcal{B}'$ and $\mathcal{S}'$ denote the sets occupied by the displaced body and its submerged part respectively; let $M' = D \iint_{\mathcal{S}'} dx \, dy$. Now Bernoulli's main problem is to find the new force of buoyancy $(0, \mathfrak{g}M')$ and the new center of buoyancy

$$\mathbf{B}' = \frac{D}{M'} \iint_{\mathcal{S}'} (x, y) \, dx \, dy. \tag{12.6}$$

Bernoulli indicates his method in an earlier paper on the statics of floating bodies to which he refers;[3] from the form of his solution it is clear that he proceeds in the present case as follows:[4] Let the displacement (12.5) be implemented through three successive transformations: First a

---

[3] Daniel Bernoulli [6]. In this paper Bernoulli treats the case of a floating body in equilibrium with an applied torque. In setting up the problem Bernoulli describes the content of our equations (12.1)–(12.6) with the help of geometrical figures. He assumes that "the nutation of the body through all intermediate positions is made through one and the same plane as that which goes through the center of gravity of the whole body and through the center of gravity of the submerged part..." and, in fact, he assumes that the body is a "vertical plane" coinciding with the plane of inclination (sec. 4, equil.). He poses the problem thus: "the hinge of the whole matter turns on determining the location of the homogeneous center of gravity of the submerged part" (sec. 8, equil.).

[4] Bernoulli gives details only for the symmetric case (in which the equilibrium position of the center of gravity lies on the perpendicular bisector of the water line).

rotation about the origin

$$T_1: (x, y) \mapsto (x, y) + \alpha(-y, x), \qquad (12.7)$$

to first order in $\alpha$; followed by a lateral translation $T_2$ by $\alpha(-G, 0)$ that puts the center of mass back on the $y$-axis;[5] and finally the vertical displacement $T_3$ by $\beta(0, 1)$. Bernoulli assumes that the water line (i.e. the intersection of $\mathscr{B}$ with the $x$-axis) is an interval, say $[-l, r]$. Under $T_1$, if for example $\alpha$, $l$, and $r$ are positive, a wedge shaped region $\mathscr{W}_r$ of $\mathscr{B}$ emerges from the fluid and Bernoulli assumes that it can be approximated (as far as its area and first moments are concerned) by the triangle $\{0 \leqslant x \leqslant r, 0 \leqslant y \leqslant x\alpha\}$ and he assumes that nothing else emerges and, similarly, he assumes that the region $\mathscr{W}_l$ that submerges can be approximated by the triangle $\{-l \leqslant x \leqslant 0, x\alpha \leqslant y \leqslant 0\}$. Thus, under $T_1$ the mass of the displaced fluid is increased by

$$-D \iint\limits_{\mathscr{W}_r} dx\, dy + D \iint\limits_{\mathscr{W}_l} dx\, dy = D \frac{l^2 - r^2}{2} \alpha. \qquad (12.8)$$

Under $T_2$ the mass is not changed; but under $T_3$, if he assumes that the region that emerges (when, say, $\beta$ is positive) can be approximated by the rectangle $\{-l \leqslant x \leqslant r, 0 \leqslant y \leqslant \beta\}$, the mass of the displaced water increases by

$$-D(r+l)\beta. \qquad (12.9)$$

Hence[6]

$$M' = M + D \frac{l^2 - r^2}{2} \alpha - D(r+l)\beta, \qquad (12.10)$$

to first order. Similarly, if $\mathscr{S}_1$ denotes the submerged part of $T_1\mathscr{B}$,

$$\frac{D}{M} \iint\limits_{\mathscr{S}_1} (x, y)\, dx\, dy = \frac{D}{M} \iint\limits_{T_1\mathscr{S}} (x, y)\, dx\, dy$$

$$-\frac{D}{M} \iint\limits_{\mathscr{W}_r} (x, y)\, dx\, dy + \frac{D}{M} \iint\limits_{\mathscr{W}_l} (x, y)\, dx\, dy$$

$$= (0, -B) + \alpha(B, 0) + \frac{D}{M} \left( \left\{ -\int_0^r x^2 \alpha\, dx - \int_{-l}^0 x^2 \alpha\, dx \right\}, 0 \right)$$

$$= \left( \left\{ B - \frac{D}{M} \frac{r^3 + l^3}{3} \right\} \alpha, -B \right) \qquad (12.11)$$

---

[5] In Bernoulli's terminology: $T_2$, for example, is described in the symmetric case as follows: if rotation is made about the mid-point of the water line, instead of about the center of gravity, then it is to be followed by a "parallel horizontal motion to restore the [center of gravity] to the vertical" (sec. 9, equil.)

[6] In Bernoulli's notation: "a portion equal to FG $\cdot \beta$ − FG $\cdot$ HN $\cdot \alpha$ emerges in the new location...," where FG is our $(r+l)$, $\beta$ is our $\beta$, HN is our $((r-l)/2)$, and $\alpha$ is our $-\alpha$; so Bernoulli's portion agrees with our $M - M'$ given in equation (12.10), with $D = 1$ (sec. 15).

to first order in $\alpha$. If $\mathscr{S}_2$ denotes the submerged part of $\mathsf{T}_2\mathsf{T}_1\mathscr{B}$, then $\mathscr{S}_2 = \mathsf{T}_2\mathscr{S}_1$ and

$$\frac{D}{M}\iint_{\mathscr{S}_2} (x, y)\, dx\, dy = \left(\left\{B - G - \frac{D}{M}\frac{r^3+l^3}{3}\right\}\alpha, -B\right). \quad (12.12)$$

Finally $\mathsf{T}_3$ causes the shift $(0, \beta)$ except for the correction

$$-\frac{D}{M}\int_{-l}^{r}\int_{0}^{\beta} (x, y)\, dx\, dy = -\left(\frac{D}{M}\frac{r^2-l^2}{2}\beta, 0\right) \quad (12.13)$$

due to the emergence of the rectangular section of the body. Thus[7]

$$\frac{D}{M}\iint_{\mathscr{S}'} (x, y)\, dx\, dy = (0, -B)+\left(\left[\left\{B - G - \frac{D}{M}\frac{r^3+l^3}{3}\right\}\alpha - \frac{D}{M}\frac{r^2-l^2}{2}\beta\right], \beta\right),$$

$$(12.14)$$

to first order in $\alpha$ and $\beta$. Finally, $\mathbf{B}'$ is obtained by multiplying this by $M/M' = 1 -(D/M)\{[(l^2 - r^2)/2]\alpha - (r + l)\beta\}+\cdots$; to first order this makes a correction only in the second component which doesn't affect the torque to be calculated below.

Bernoulli has now that the total force due to the forces of gravity and buoyancy acting on the displaced body is[8]

$$(0, \mathfrak{g}M' - \mathfrak{g}M) = \left(0, \mathfrak{g}D\left\{\frac{l^2-r^2}{2}\alpha - (r+l)\beta\right\}\right) \quad (12.15)$$

to first order by (12.10) and he has that the total torque due to these forces about the center of mass is[9]

$$\{\mathbf{B}' - \mathbf{G}'\}\wedge (0, \mathfrak{g}M') = \mathfrak{g}M\left[\left\{B - G - \frac{D}{M}\frac{r^3+l^3}{3}\right\}\alpha - \frac{D}{M}\frac{r^2-l^2}{2}\beta\right] \quad (12.16)$$

to first order, by (12.14).

---

[7] In Bernoulli's notation: the horizontal displacement $bc$ of the center of buoyancy is

$$\frac{(\mathsf{AB}\cdot M +\frac{1}{3}\mathsf{FN}^3+\frac{1}{3}\mathsf{GN}^3)\alpha}{M} - \frac{\mathsf{FG}\cdot\mathsf{HN}\cdot\beta}{M},$$

where $\mathsf{AB}$ is our $(G - B)$, $\mathsf{FN}$ is our $l$, $\mathsf{GN}$ is our $r$, and the other symbols are as given in note 6. This is the same as the horizontal coordinate in (12.14), with $D = 1$ (sec. 15).

[8] In Bernoulli's notation: "...the vertical force $\pi$ is equal to the weight of the portion [which emerges], whence there is immediately $\pi = \mathsf{FG}\cdot\beta - \mathsf{FG}\cdot\mathsf{HN}\cdot\alpha$," the weight of the portion given above in note 6 (with $\mathfrak{g} = 1$).

[9] Bernoulli expresses the torque in terms of a pair of equal and opposite horizontal forces $P$ acting at the center of mass and at a distance $\mathsf{AR}$, equal to $\sqrt{I_0/M}$, from the center of mass (see the last paragraph of Chapter 12, section 1), with units chosen so that this distance is equal to one. He writes the force as the total mass multiplied by the horizontal displacement $bc$, given in note 7, which gives the same result as equation (12.16).

Bernoulli makes the usual assumption that each mass element $\rho \, dx \, dy$ of the body is accelerated by an harmonic force proportional to its displacement (12.5), namely

$$-\mathfrak{a}^{-1}\{\alpha(-y-G, x)+\beta(0, 1)\}\rho \, dx \, dy. \tag{12.17}$$

The pendulum condition, as it had been used by Euler (Chapter 10) as an equilibrium condition, asserts that the total force and total torque about the center of mass due to the forces (12.17) equal the total force and torque of (12.15) and (12.16) respectively. The total force of (12.17) is

$$-\mathfrak{a}^{-1}\iint_{\mathcal{B}} \{\alpha(-y-G, x)+\beta(0, 1)\}\rho \, dx \, dy = -\mathfrak{a}^{-1}M\beta(0, 1) \tag{12.18}$$

by (12.1) and (12.2) and the total torque is

$$-\mathfrak{a}^{-1}\iint_{\mathcal{B}} (x-\{y+G\}\alpha, \{y+G\}+x\alpha) \wedge (-\{y+G\}\alpha, x\alpha+\beta)\rho \, dx \, dy$$

$$= -\mathfrak{a}^{-1}\iint_{\mathcal{B}} \{x^2+\{y+G\}^2\}\alpha\rho \, dx \, dy$$

$$= -\mathfrak{a}^{-1}I\alpha \tag{12.19}$$

to first order, by (12.2), where $I$ is the moment of inertia of the body about its center of mass. Thus, the pendulum condition is the pair of equations[10]

$$\begin{cases} -\mathfrak{a}^{-1}M\beta = \mathfrak{g}D\left\{\dfrac{l^2-r^2}{2}\alpha-(r+l)\beta\right\} \\[2mm] -\mathfrak{a}^{-1}I\alpha = \mathfrak{g}M\left[\left\{B-G-\dfrac{D}{M}\dfrac{r^3+l^3}{3}\right\}\alpha-\dfrac{D}{M}\dfrac{r^2-l^2}{2}\beta\right]. \end{cases} \tag{12.20}$$

Finally Bernoulli obtains the two simple modes by solving (12.20). He considers separately the case $r = l$ in which the center of mass lies on the bisector of the water line. In this case the equations (12.20) separate and one mode consists of pure rocking and the other mode of pure bobbing. In the asymmetric case, $r \neq l$, he solves (12.20) for the bob per rock, $\beta/\alpha$, of the two modes and the corresponding periods, or rather lengths of the isochronous pendula $\mathfrak{a}\mathfrak{g}$. He makes considerable note of the fact that the simple modes thus involve both bobbing and rocking in specific proportions:

In the [symmetric] case, during its change of place the plane [i.e. body] simply rotates about its center of gravity and as soon as the forces cease to act the plane

---

[10] Bernoulli works with the ratio of these equations. He writes that "it is necessary that the ratio of the accelerating forces $P/M$ and $\pi/M$ be the same as that of the paths described, $\alpha$ and $\beta$, whence there arises the equation $\alpha\pi = \beta P$" and he substitutes the expressions for $P$ and $\pi$ that are given in notes 8 and 7 above (sec. 17).

will begin to be agitated about the center of gravity with a reciprocal motion. In the other case, when the forces cease, the plane will be agitated with a mixed motion, the one oscillatory and rotational about the center of gravity, the other being a vertical and parallel motion in which the plane alternately ascends and descends.

We should note that Bernoulli doesn't use the moment of inertia $I$ per se; he uses the length $\sqrt{I/M}$ of whose properties he is enamored (e.g. if the body is hung in a gravitational field at a point that distance from its center of mass, then its oscillations as a compound pendulum will have a shorter period than they would were it hung from any other point). Rather than use torque he uses force exerted on an arm of that length (balanced by a force at the fulcrum) and he has the force act on a point mass $M$ (which is appropriate since a mass $M$ at that distance has moment of inertia $I$ about the fulcrum). He applies the pendulum condition directly to the mass point $M$.

## 12.2. Remarks

Bernoulli seems to have been alone in considering the asymmetric case where both rocking and bobbing occur in the simple modes. The symmetric case, since it reduces immediately to two problems each in one degree of freedom, does not really fall into our topic. But we should mention that it received attention by others. The familiar notion of the "metacentric height," which in our notation, for two dimensions and small angles of heel, is

$$\mathfrak{m} = G - B + \frac{D}{M}\frac{2r^3}{3}, \qquad (12.21)$$

seems to have been introduced by Bouguer in his *Traité du navire*[11] of 1746. In this notation, the torque (12.16) is $-\mathfrak{g}M\mathfrak{m}\alpha$ and the "metacenter," fixed in the body a distance $\mathfrak{m}$ above the center of mass, remains always above the center of buoyancy during rocking. It seems that Bouguer's work was independent. Euler may have been the first to obtain the torque (12.16) (in the $r = l$ case). At any rate, in 1735, Euler reported to the St. Petersburg Academy on a letter from a Frenchman, named de la Croix, who seems to have suggested that the restoring torque on a floating body should be calculated on the basis of Archimedes' principle;[12] and Euler asserted that he had already obtained the torque in question and he gave it in the form

---

[11] Stoot [1].

[12] Euler [10]. This report and the second, Euler [11], given the following year, were not published at the time.

(12.16) (for $r = l$). In his report, Euler gave as an application the result that a homogeneous cube will float stably in water, with its sides parallel or vertical to the surface, if its specific density is less than $\frac{1}{2} - \sqrt{3}/6$ or greater than $\frac{1}{2} + \sqrt{3}/6$ (and less than one). A year later he gave a second report on a letter form de la Croix giving more details. At about this time Euler began working out various examples of stability of solid bodies floating in water that he collected in his book *Scientia Navalis* of 1749. Euler began a correspondence with Johann Bernoulli on his work on floating bodies[13] on 10 December 1737 and he gave rough details on 30 July 1738, suggesting that Johann Bernoulli communicate the formulas to Daniel Bernoulli.[14] At the beginning of his paper on the statics of floating bodies, Daniel Bernoulli acknowledges his indebtedness to Euler for suggesting the method. What is interesting, though, is that in spite of his pre-eminence in the field, Euler did not understand Daniel Bernoulli's analysis of the asymmetric case. In fact, Euler proved himself quite confused in a letter[15] to Johann Bernoulli of 19 January 1740 in which he wrote that since the motion of the center of mass depends on the distance between the vertical lines drawn through the center of mass of the submerged part and that of the whole ship (there being no motion when the lines coincide), it is impossible for a boat to undergo "pure horizontal vibrations, but they will perpetually be combined with vertical oscillations. . . . , which follow their own law, so that unless the times of both [types of] oscillations are equal or commensurate, regular total oscillations are lacking." Euler added that Daniel, in disagreement with him, had decided that "both oscillations must always compose themselves to uniformity, so that they appear as a single kind of oscillation."

In a letter of 7 March 1739, Daniel Bernoulli had already written to Euler that the oscillations in the asymmetric case were a delicate matter in which one could easily take *nubem pro Junone*. In particular, he says that irregular oscillations are possible (i.e. superpositions of the two modes, though not recognized as such).[16]

## 12.3. Dangling Rod

Bernoulli considers the small oscillations of a rigid rod hanging at the end of a massless string. Let the gravitational field act in the direction $(-1, 0)$ in the $z$–$y$ plane; let the string have length $l$ with one end fixed at $(l, 0)$; and let the rod, attached to the other end of the string, have length

---

[13] Eneström [1], V, 270–271.

[14] *ibid.* V, 278–282.

[15] *ibid.* VI, 48.

[16] Fuss [1], II, p. 455.

$L$. Thus, in equilibrium, the system hangs along the $z$-axis from $l$ down to
$-L$. For small displacements from equilibrium, the configuration of the
system is described by the equation

$$y = \alpha(z - k), \qquad -L \leqslant z \leqslant 0, \tag{12.22}$$

where $y = y(z)$ is the displacement of the rod element $\rho(z)\,dz$ from its
equilibrium position, $\alpha$ is the angle which the rod makes with the $z$-axis,
and $k$ is the point on the $z$-axis at which the line of the rod intersects.
Now, Bernoulli uses $\alpha$ and $k$ to parametrize the configuration. The parameter $\alpha$ is similar to those of all preceding problems; but $k$ is singular at
the equilibrium point. Bernoulli's use of a singular parameter gives the
problem a new appearance. The parameter $k$ has a priori the property that
it is fixed in the motion of a simple mode (since a simple mode involves
only scaling in time of the displacement).

Bernoulli's method for dealing with the small oscillations of a rigid body
should be clear already: He imposes the pendulum condition by equating
respectively the total force and torque due to the applied forces with the
total force and torque of the harmonic forces. In the case at hand, he
calculates torques about the end point $(0, y(0))$. The total force is clearly
horizontal and due to the string alone which, because the oscillations are
small, is at tension $\mathfrak{g}M$, where $M = \int_{-L}^{0} \rho\,dz$ is the mass of the rod. This
force is[17]

$$-\mathfrak{g}M\frac{y(0)}{l} = \mathfrak{g}M\frac{k}{l}\alpha \tag{12.23}$$

by (12.22). The total torque about the end at which the string is attached
is due to gravity alone; it is

$$\int_{-L}^{0} (z, y - y(0)) \wedge (-\mathfrak{g}\rho, 0)\,dz = -\mathfrak{g}MG\alpha \tag{12.24}$$

by (12.22), with $-G$ the center of mass of the rod along the $z$-axis.[18] Next,
the harmonic force on $\rho\,dz$ is[19]

$$(0, -\mathfrak{a}^{-1}\rho\,dz\,y) = (0, -\mathfrak{a}^{-1}\rho\,dz\,\alpha\{z - k\}). \tag{12.25}$$

---

[17] In Bernoulli's notation: the force is expressed in terms of two angles—our $\alpha$, and the
angle between the line of the rod and the string. Thus there are two terms, one which Bernoulli
characterizes as being "due to the oblique action of the string," and another, "due to the
weight of the rod:"

$$\frac{(n - \alpha)q}{n(\alpha + g)}M - \frac{q}{\alpha + q}M,$$

where $n$ is our $l$, $\alpha$ is our $k$, $g$ is our $G$, $q$ is our $-\alpha(G + k)$, and $\mathfrak{g} = 1$.

[18] In Bernoulli's notation: the torque, which he gives directly, is $(Mq/(\alpha + g))g$.

[19] In Bernoulli's notation: the negative of the *harmonic force is* $x\,d\xi/a$, where $x$ is our
$(k - z)$, $d\xi$ is our $-\rho\,dz$, and $a$ is our $-\mathfrak{a}/\alpha$ (sec. 9).

The total force due to (12.25) is horizontal and given by[20]

$$-a^{-1}\alpha \int_{-L}^{0} \{z-k\}\rho \, dz = a^{-1}M\{G+k\}\alpha, \qquad (12.26)$$

while the total torque about $(0, y(0))$ is[21]

$$\int_{-L}^{0} (z, y-y(0)) \wedge (0, -a^{-1}\rho\alpha\{z-k\}) \, dz = -a^{-1}\alpha \int_{-L}^{0} z\{z-k\}\rho \, dz$$

$$= -a^{-1}\{I_0 + MkG\}\alpha, \qquad (12.27)$$

where $I_0$ is the moment of inertia about the end point. Equating (12.23) and (12.26), Bernoulli has[22]

$$ag = l\frac{G+k}{k}, \qquad (12.28)$$

and equating (12.24) and (12.27), he has[23]

$$ag = \frac{I_0}{MG} + k. \qquad (12.29)$$

The equations (12.28) and (12.29) have a certain eighteenth century elegance about them. For example, as Bernoulli notes, (12.29) says that if the dangling rod is oscillating in a simple mode then a simple pendulum of the same frequency will have length equal to the length it would have were the rod to be simply swinging about its end point plus (or minus) the distance to the point about which the rod is actually oscillating. In Chapter 14 we will come to Euler's interpretation of (12.28).

Bernoulli solves the equations (12.28) and (12.29) to obtain the values of $k$ for the two simple modes and the corresponding lengths $ag$. Finally, he points out that the solution generalizes to the case of an arbitrary (two dimensional) body suspended at any point from the string since the relevant quantities are just $M$, $G$, and $I_0$ which are defined generally, relative to the point of suspension. In connection with this he works out a number of examples; but we will not go through these details.

## 12.4. Remarks on Superposition

One almost expects Bernoulli to discover the principle of superposition in these papers—especially since in the case of a body dangling from a

---

[20] In Bernoulli's notation: $\int x \, d\xi = ([\alpha]+g)M$.

[21] In Bernoulli's notation: $\int ((x-\alpha)x \, d\xi/a) = (gM/a)(\alpha+L)$, where $L$, "the length of the simple pendulum isochronous with the minimal oscillations of the rod CH suspended from C," is our $I_0/MG$.

[22] This is Bernoulli's equation I, $n(\alpha+g)^2 = a\alpha q$.

[23] This is Bernoulli's equation II, $(\alpha+1)(\alpha+g) = aq$.

rather long string, the two modes are visually easily separable even when superimposed. One's expectations may be further encouraged by the title of the dangling rod paper, "De oscillationibus compositis. . . . ," which seems to refer to "composite oscillations." But he does not. He believes that small oscillations are either isochronous and regular or that they are irregular; furthermore he believes that the irregular ones will settle down to simple modes for undescribed reasons. His word "*compositis*" should be translated as "composed." It refers to the idea that a system in many degrees of freedom "composes" its configuration in a particular way when it oscillates. In the case of a simple mode, it "composes" itself into a particular shape that changes only by scaling in oscillating. He mentions for comparison the case of isochronous vibrations of the string. He clearly considers Taylor's analysis, generalized to allow higher modes, to be sufficient. Since non-isochronous vibrations are not heard he has Taylor's paradox and he follows Taylor in believing the solution to be that the vibrations that are not in a simple mode quickly decay to a simple mode. If Bernoulli were thinking of string vibrations that are not in a simple mode as superpositions, he would not conclude that they must be irregular and not isochronous (since he is perfectly aware that the higher modes in this case are harmonics). Also in the introduction to the paper on the dangling rod, Bernoulli refers to the paper on floating bodies for an example in which "two kinds of oscillations" exist simultaneously. Read casually, this may seem like superposition; but it is clear from section 12.1 above that he is referring to the fact that both rocking and bobbing are involved in the simple modes. Bernoulli's most tantalizing statement comes in his paper on floating bodies: "These remarks serve for understanding the tremulous motions of sonorous strings: for the sound of one and the same string can be composed from many tones." Yet the remarks in question refer only to simple modes; so it is clear that he is referring only to separately sounding overtones.

# 13. Daniel Bernoulli (1742)

Daniel Bernoulli had written Euler in 1735 of his work on the (transversally) vibrating rod (see p. 70). Around 1740, he finally wrote a paper on the subject.[1] In this work, he obtains the pendulum condition as $\hat{a}\, d^4 y/dz^4 = y$, and the boundary conditions appropriate for the rod clamped at one end; he solves it approximately by both the series method of undetermined coefficients and in the form[2]

$$y = A \cosh rz + B \sinh rz + C \cos rz + D \sin rz, \qquad \hat{a} = 1/r^4, \quad (13.1)$$

for the fundamental solution; in the latter form he finds approximate values of $r$ for the higher modes; he effectively obtains the flexural rigidity $e$ (where $\hat{a} = ae/p$) in terms of the deflection of the rod under a certain force at the free end; thus he predicts the fundamental frequency of a needle, clamped at one end, which he confirms experimentally. After completing this work, Bernoulli took up the problem of the vibrating rod that is free at both ends, apparently inspired to understand the vibrations of carillon rods. He wrote a second paper[3] that treats this case in detail, calculating both nodal points and frequencies for the first five modes. He reports on experiments in which the different modes are induced by gently holding a rod at predicted nodal points and heard to vibrate at predicted frequencies. In doing these experiments, Bernoulli hears superimposed modes and he gives a theoretical argument that superposition can occur. (He does not, however, give a Superposition Principle; that is, he doesn't assert that all small vibrations are superpositions of simple modes.) Both papers appeared in 1751 in the volume for 1741–1743 of the St. Petersburg Academy. We will discuss the two papers together, covering the basic theory of the vibrating rod in section 13.1, the predictions and experiments on absolute frequency in section 13.2, and superposition in section 13.3.

---

[1] Daniel Bernoulli [9].
[2] In Bernoulli's notation:

$$y = ae^{x/f} + be^{-x/f} + h \sin \text{arc}\, (x/f + n),$$

where $x$ is our $z$, $f$ is our $1/r$, and "sin arc" means "sine."
[3] Daniel Bernoulli [10].

## 13.1. Vibrating Rod

We will use the notation that we used in discussing Euler's 1735 paper (Chapter 10). The unstressed rod lies along the $z$-axis of the $z$-$y$ plane: If the rod is clamped at one end, it will be supposed to lie along the interval $0 \leqslant z \leqslant l$ and be free at the origin; if it is free at both ends it will be supposed to lie along the interval $-l/2 \leqslant z \leqslant l/2$. Bernoulli considers only small deflections to a curve $y(z)$ so that the curvature at $z$ is $d^2y/dz^2$. He assumes as a common hypothesis that in equilibrium the bending moment at $z$ is proportional to the curvature and we will again denote the constant of proportionality (the flexural rigidity) by $e$.

If $p(z)$ is the force density acting in the $y$-direction, Bernoulli calculates its contribution to the bending moment as follows: As before we will consider the rod to the left of $z$ and calculate the negative of the moment due to $p$ for the case of the rod having its free end at the origin. The total force is $\int_0^z p\, dz$ and Bernoulli considers it as acting at the "center of force,"[4]

$$\xi = \int_0^z \hat{z} p\, d\hat{z} \Big/ \int_0^z p\, d\hat{z}. \tag{13.2}$$

Thus, Bernoulli writes the negative of the moment at $z$ due to $p$ as

$$(z - \xi) \int_0^z p\, dz, \tag{13.3}$$

which is the length of a simple moment arm times the total force. (This isn't defined at those $z$ for which the total force is zero; but these points will be isolated.) Bernoulli regards the notion of a "center of force" as a trivial generalization of the notion of a center of gravity when the force density can be given as a scalar $p$ (times a fixed vector). In his earlier paper on the bending of a beam under its own weight,[5] Bernoulli had been concerned with the gravitational force. Here he considers only the negative of the harmonic force $p = \mathfrak{a}^{-1}\rho y(z)$, so that the equilibrium condition is the pendulum condition[6]

$$\mathfrak{a}e \frac{d^2y}{dz^2} = (z - \xi) \int_0^z \rho y\, dz. \tag{13.4}$$

This is Bernoulli's basic equation, with $\xi$ defined by (13.2) with $y$ in the place of $p$. Bernoulli differentiates (13.4) and then uses (13.2), or rather

---

[4] In Bernoulli's notation: the derivative of equation (13.2) is $y\, dx(x - \xi) = d\xi \int y\, dx$, where $x$ is our $z$, in the case of $y$ in the place of $p$ (first paper, sec. 4).

[5] Daniel Bernoulli [1].

[6] In Bernoulli's notation:

$$\frac{d\, dy}{d[x]} = \frac{dx}{m^4}(x - \xi) \frac{G}{c} \int y\, dx,$$

where $m^4$ is our $e$, and $G/c$ is our $\mathfrak{a}^{-1}\rho$, (Bernoulli's $\rho$ is constant; Bernoulli's $G/c$ will be explained in section 13.2).

its derivative, to eliminate $\xi$ and $d\xi/dz$, obtaining

$$ae \frac{d^3 y}{dz^3} = \int_0^z \rho y \, dz \tag{13.5}$$

and, by a final differentiation,[7]

$$\frac{1}{r^4} \frac{d^4 y}{dz^4} = y, \quad \text{where } \frac{1}{r^4} = \frac{ae}{\rho}. \tag{13.6}$$

At this point Bernoulli explains in detail that the length of the simple isochronous pendulum is $a\mathfrak{g}$.

Bernoulli notes that the boundary conditions at the clamped end are

$$y(l) = 0 = \frac{dy}{dz}\bigg]_{z=l} \tag{13.7}$$

and, from (13.4) and (13.5), that at the free end they are

$$\frac{d^2 y}{dz^2}\bigg]_{z=0} = 0 = \frac{d^3 y}{dz^3}\bigg]_{z=0}. \tag{13.8}$$

In his second paper on the free–free case he has, similarly, the boundary conditions

$$\frac{d^2 y}{dz^2}\bigg]_{z=\pm l/2} = 0 = \frac{d^3 y}{dz^3}\bigg]_{z=\pm l/2}. \tag{13.9}$$

Bernoulli first approaches the solution of (13.6), subject to (13.7) and (13.8), or (13.9), by the series method of undetermined coefficients exactly as Euler had in his 1735 paper (Chapter 10). His method for finding roots is presumably similar to that that he used in the problem of the hanging chain. In the first paper he says that he generally prefers the series method for its ease in calculations (see note 13); but in the second paper he says that the calculations are troublesome and for higher modes, insurmountable. He gives the following approximate results: For the free–clamped case, $(rl)^4 = 12.25$; for the free–free case, $(rl)^4 = 496$; 3,824; and 14,608. (Correct values would be respectively 12.3623; 500.564; 3803.54; and 14,617.6).

Bernoulli's second approach is to seek the solutions in the form[8] (13.1), where $r$ and the relative coefficients are to be determined from the boundary

---

[7] In Bernoulli's notation: $d^4 y = y \, dx^4/f^4$, where $1/f^4 = G/m^4 c$ (thus Bernoulli's $1/f$ is our $r$) (first paper, secs. 4, 5, 7).

[8] Bernoulli writes that he would not have attempted to do this had he not understood from Euler that it was possible (first paper, sec. 6). Euler discovered the general solution in terms of sines and exponentials for linear equations with constant coefficients in 1739, although not for the case of repeated roots. He mentioned his solution in a letter to Johann Bernoulli without reference to the vibrating rod, but he gave equation (13.6) as his first example. A little later Daniel Bernoulli gave the general solution with the modification needed for repeated roots. See Truesdell [1], p. 177.

conditions (13.7) and (13.8), or (13.9). He does not unite the two methods (e.g. by expanding the terms in (13.1) into series); he checks agreement only by comparing numerical solutions.

For the free–clamped case, Bernoulli has, by substituting (13.1) into (13.8), that $A = C$ and $B = D$ and whence, by substituting (13.1) into (13.7), that

$$\begin{cases} A \{\cosh rl + \cos rl\} + B \{\sinh rl + \sin rl\} = 0 \\ A \{\sinh rl - \sin rl\} + B \{\cosh rl + \cos rl\} = 0 \end{cases} \tag{13.10}$$

and, by taking ratios, the equation[9] for $rl$: $\{\cosh rl + \cos rl\}^2 = \sinh^2 rl - \sin^2 rl$ or

$$\cosh rl \cos rl = -1, \tag{13.11}$$

which has the asymptotic approximation $\cos rl \approx 0$ so that[10]

$$rl \approx N\pi/2, \qquad N = 1, 3, 5, 7, \ldots \tag{13.12}$$

are approximate solutions.

For the free–free case, Bernoulli first decides on intuitive grounds that the solutions are either even, i.e. $B = 0 = D$, or odd, i.e. $A = 0 = C$, and that there is no solution with a single node (though of course these things are consequences of the boundary conditions). Then on substituting (13.1) into (13.9), Bernoulli has for the even case,

$$\begin{cases} A \cosh \dfrac{rl}{2} - C \cos \dfrac{rl}{2} = 0 \\[2mm] A \sinh \dfrac{rl}{2} + C \sin \dfrac{rl}{2} = 0 \end{cases} \tag{13.13}$$

and for the odd case,

$$\begin{cases} B \sinh \dfrac{rl}{2} - D \sin \dfrac{rl}{2} = 0 \\[2mm] B \cosh \dfrac{rl}{2} - D \cos \dfrac{rl}{2} = 0 \end{cases} \tag{13.14}$$

---

[9] In Bernoulli's notation:

$$\frac{l}{f} = 2 \arcsin \frac{1 + e^{l/f}}{\pm\sqrt{(2 + 2e^{2l/f})}}$$

(first paper, secs. 10, 17).

[10] In Bernoulli's notation: $(2r - 1)q$, where $r = 1, 2, 3, \ldots$, and $q$ is $\pi/2$.

and, by taking ratios, the equation[11] for $rl$:

$$\tanh \frac{rl}{2} = \pm\tan \frac{rl}{2}, \tag{13.15}$$

where the minus and plus signs correspond respectively to the even and odd cases. Asymptotically, (13.15) becomes $\tan (rl/2) \approx \pm 1$, so that[12]

$$rl \approx N\pi/2, \qquad N = 3, 5, 7, \ldots \tag{13.16}$$

are approximate solutions to (13.15). (There is no solution corresponding to $N = 1$). $N = 3, 7, 11, \ldots$ corresponds to the even solutions, etc. and Bernoulli notes that $(N+1)/2$ is the corresponding number of nodal points.

Bernoulli notes that the approximate solutions (13.12) and (13.16) are excellent for the higher modes, $N \geq 5$. His method for improving this approximation is to substitute $rl = N\pi/2 + \varepsilon$ into (13.11) or (13.15), expand to second or higher order in $\varepsilon$, and solve the resultant quadratic equation.[13] For the first mode in the free–clamped case he obtains $1/rl = 0.535$, which agrees with his previously obtained result $\sqrt{\frac{2}{7}} = 0.53452\ldots$ (The correct value would be 0.533305.) For the first mode in the free–free case he obtains $rl = 4.7213$, which agrees with his previously obtained result $(497)^{1/4} = 4.7192\ldots$ (The correct value would be 4.73004.) Let us set out these results: Since $\mathfrak{a} = \rho/r^4 e$,

$$\mathfrak{a} = (rl)^{-4} l^4 \rho/e \tag{13.17}$$

where the coefficient $(rl)^{-4}$ is determined from the boundary conditions, for each mode, $l$ is the length and $\rho/e$ is characteristic of the rod. Since $\nu = 1/2\pi\sqrt{\mathfrak{a}}$, (13.17) gives the frequencies

$$\nu = \frac{(rl)^2}{2\pi} \frac{1}{l^2} \sqrt{e/\rho}. \tag{13.18}$$

Corrected values for $(rl)^2$ are: 1) in the free–clamped case, 3.5160; 22.034; and $N^2\pi^2/4$, $N = 5, 7, 9, \ldots$ for the first, second, and higher modes and

---

[11] In Bernoulli's notation: for the odd case (the "second class"),

$$\frac{\sin \text{ arc } l/2f}{e^{l/2f} - e^{-l/2f}} = \frac{\cos \text{ arc } l/2f}{e^{l/2f} + e^{-l/2f}},$$

in the even case (the "first class") his expression has the wrong sign (second paper, sec. 14).

[12] In Bernoulli's notation: $l/f = mq$, where $m$ is our $N$ and $q$ is $\pi/2$.

[13] For the substitution into equation (13.15), Bernoulli gives the details for the case $N = 3$ (second paper, sec. 14). Here he expands to second order in $\varepsilon$ (his $2\alpha$). For the substitution into equation (13.11) Bernoulli gives only the result (first paper, sec. 10). In this case, an expansion to second order is insufficient for the fundamental mode ($N = 1$). Bernoulli presumably expanded at least to third order and then used his root technique (of Chapter 9, section 4). This is more tedious than Bernoulli's first method (in which he treats the solution itself as a series).

2) in the free–free case, 22.373; and $N^2\pi^2/4$, $N = 5, 7, 9, \ldots$, for the first and higher modes respectively.

For the case of the vibrating rod that is free at both ends, Bernoulli calculates nodal points[14] to a good approximation. For this he uses the approximations

$$\frac{rl}{2} \approx N\pi/4, \qquad N = 3, 5, 7, \ldots, \tag{13.19}$$

(even for the fundamental mode, $N = 3$). With (13.19), by (13.13) and (13.14), if $C = D = 1$, (13.1) becomes

$$y = \frac{\cos\dfrac{N\pi}{4}}{\cosh\dfrac{N\pi}{4}}\frac{\cosh}{\sinh}\left(\frac{N\pi}{2}\frac{z}{l}\right) + \frac{\cos}{\sin}\left(\frac{N\pi}{2}\frac{z}{l}\right) \tag{13.20}$$

where cosh and cos are taken in the even cases, $N = 3, 7, 11, \ldots$, and sinh and sin are taken in the odd cases, $N = 5, 9, 13, \ldots$. Since $\cos(N\pi/4)/\cosh(N\pi/4)$ is small, the nodal points $z/l$ at which $y$ is zero are roughly equal to $n/N$, $|n| < N/2$ with $n$ odd in the even case and even in the odd case:

$$\frac{z}{l} = \frac{n}{N} + \frac{2}{N\pi}\varepsilon \tag{13.21}$$

where $\varepsilon$ is small. Following Bernoulli, we substitute (13.21) into (13.20) with $y = 0$ to obtain

$$0 = \frac{\cos\dfrac{N\pi}{4}}{\cosh\dfrac{N\pi}{4}}\frac{\cosh}{\sinh}\left(\frac{n\pi}{2} + \varepsilon\right) + \left\{\cos\frac{n\pi}{2} - \sin\frac{n\pi}{2}\right\}\sin\varepsilon. \tag{13.22}$$

But to solve this for $\varepsilon$, Bernoulli makes the approximations

$$\sin\varepsilon \approx \varepsilon \quad \text{and} \quad \frac{\cosh}{\sinh}\left(\frac{n\pi}{2} + \varepsilon\right) \approx \frac{1}{2}\{1 + \varepsilon\}\exp\frac{n\pi}{2} \tag{13.23}$$

since it suffices to consider only $n > 0$. With these approximations, he solves (13.22) for $\varepsilon$:

$$\varepsilon \approx \left[\left\{\sin\frac{n\pi}{2} - \cos\frac{n\pi}{2}\right\}\left\{\cos\frac{N\pi}{4}\right\}^{-1}\exp\left(\frac{N - 2n}{4}\pi\right) - 1\right]^{-1}$$

$$= \left[(\pm 1)(\pm\sqrt{2})\exp\left(\frac{N - 2n}{4}\pi\right) - 1\right]^{-1}. \tag{13.24}$$

---

[14] Bernoulli uses Sauveur's expression "node," which he had not used in his earlier papers.

Bernoulli makes an adequate evaluation of (13.24) for the first few modes to obtain the approximate nodal points via (13.21). We will list a little more accurately the values of $z/l$ in (13.21) that follow from (13.24) together with some correct nodal points:

| $N$ | $n$ | $\dfrac{n}{N}+\dfrac{2}{N\pi}\varepsilon$ | correct node $z/l$ | |
|---|---|---|---|---|
| 3 | 1 | 0.2816 | 0.2758 | |
| 5 | 2 | 0.3690 | 0.3679 | |
| 7 | 1 | 0.1441 | 0.1442 | |
| 7 | 3 | 0.4064 | 0.4056 | |
| 9 | 2 | 0.2232 | | (13.25) |
| 9 | 4 | 0.4272 | | |
| 11 | 1 | 0.0909 | | |
| 11 | 3 | 0.2735 | | |
| 11 | 5 | 0.4404 | | |

(The negatives are also nodes and in the odd cases, $N = 5, 9, \ldots$ zero is also a node.)

## 13.2. Absolute Frequency and Experiments

Now Bernoulli gives an independent experimental method for determining $\sqrt{e/\rho}$. His discussion of this occurs rather early in the first paper, in the midst of his other analysis, because he is concerned to use directly measurable quantities. Bernoulli considers the rod clamped at one end and held in static equilibrium by a perpendicular force $P$ at the other end. He asserts that one should follow the same reasoning that led to equation (13.6). Thus, the force density $p(z)$ that occurs in equation (13.2) and (13.3) is concentrated at the origin and the total force $\int p\, dz = P$; so in the place of (13.3) one has $zP$ for the bending moment at $z$, again set equal to $e\, d^2y/dz^2$ for equilibrium:

$$e\frac{d^2y}{dz^2} = zP, \tag{13.26}$$

to which one adds the conditions $y(l) = 0 = dy/dz]_{z=l}$. Bernoulli solves this, obtaining

$$y = \frac{P}{6e}z^3 - \frac{P}{2e}l^2z + \frac{P}{3e}l^3. \tag{13.27}$$

The constant term is the deflection at the free end, $\delta$; thus

$$e = \frac{l^3 P}{3\delta},$$
(13.28)

where $l$, $P$ and $\delta$ can be measured in a simple experiment. Of course if the deflection were too large, $d^2 y/dz^2$ would be incorrect as the curvature; but it is experimentally easy to see when $\delta$ ceases to be proportional to $P$. As Euler pointed out, however, one obtains a more accurate measurement of the flexural rigidity $e$ from (13.18) if one measures the frequency.[15] Still, there is something impressive about Bernoulli's prediction of frequency from static measurements which thus parallels Taylor's determination of the frequency of a vibrating string. For the rod clamped at the same length for both the deflection experiment and the vibration, Bernoulli puts the formulas (13.28) and (13.17) together to obtain[16]

$$\mathfrak{a} = \frac{3\rho\delta l}{(rl)^4 P}.$$
(13.29)

Concerning these results, Bernoulli writes:[17]

As far as I know, nobody yet has given clearly a discussion about the sounds that elastic lamina vibrating in the described way produce or has at least given some observations of the kind of which many had been given for the sounds of tense musical strings before a complete theory of them was found by Taylor, my father, and others. It is clear that before this theory was extracted by analytic calculation, it had already been established by observations that the number of vibrations in a tense string in a given time is in a proportion composed of 1) the reciprocals of the lengths of the strings, 2) the reciprocals of the weights of the strings, and 3) the square roots of the weights stretching the strings: to define that number absolutely from these given [quantities], that was left for the sublime theory—, although [the frequencies] were found by many artifices from a long time since, so that they were found also by experiments and indeed with such success that the results did not deviate much from the truth. But finally the matter was advanced from that state so that knowing the weight of the string, the weight stretching the string, and the length of the string, the true number of vibrations in a given time could be obtained. ... We are now drawing out for the lamina those things that have already been found for the string. ...

In a letter to Euler of 20 September 1741, Bernoulli wrote:[18]

For some time now I have devoted most of my time to calculating the diverse pitches and other properties of elastic lamina which has given me the opportunity for many beautiful, entirely new experiments (that agree perfectly with my theory).

---

[15] Euler [5], par. 40.

[16] In Bernoulli's notation: for the case $rl = \sqrt{\frac{7}{2}}$, $L = (\frac{12}{49}p/P)C$, where $L$ is our $\mathfrak{a}\mathfrak{q}$, $p$ is our $\rho l$, and $C$ is our $\delta$ (first paper, sec. 9) and, in general, $L = (3p/P)(f^4/l^4)C$ (second paper, sec. 5).

[17] Daniel Bernoulli [7], sec. 13.

[18] Fuss [1], II, pp. 477–478.

We presume that Bernoulli was writing the first paper at about this time.[19] The main experiment that he discusses in this paper concerns a needle of total length 0.327 Rhenish Feet[20] (10.26 cm), weight 17 grains (1.11 grams), free length when clamped in a "wall," 0.297 Rhenish Feet (9.32 cm), and when a force of 1000 grains (65.3 grams) is applied at its free end it deflects by $\delta = 0.027\ RF$ (0.847 cm). Since the acceleration of gravity is $g = \pi^2 3.166\ RF$ per sec$^2$, $\rho = 1.664$ grains sec$^2$ per $RF^2$. With these values and $(rl) = \sqrt{\frac{7}{2}}$ substituted into (13.29) one obtains $a = 3.27 \cdot 10^{-6}$ sec$^2$ or a frequency of about 88 cycles per sec.[21] Bernoulli checks the pitch of the needle's vibration with a tone of this frequency and finds agreement. (We did a similar experiment very casually; in our case $\sqrt{e/\rho}$ was 1.21 times the value in Bernoulli's case and the other parameters were likewise similar. For us, the predicted frequency agreed to within a whole tone.[22] We bowed the needle to obtain a sustained pitch. In this way we were also able to obtain two higher modes whose frequencies were consistent to within a quarter tone with the fundamental, according to (13.18).)

Bernoulli also describes experimental confirmation that the frequency is proportional to $1/l^2$ as given in (13.18). This he does by shortening the needle by $1/\sqrt{2}$ of its length to get the octave. He goes on to write that

The instrument that is called a carillon[23] in French follows this law, since it is made of steel rods differing only in length, and the rods are used also in domestic clocks, even though the rods do not perform their minimal vibrations [as free-clamped rods] but are free: however I intend to follow this other kind of vibration [i.e. the free–free case] on another occasion.

The carillon rods that Bernoulli observed actually dangled from strings; and Bernoulli begins his second paper with the problem of the motion of a dangling rod that is struck. He decides that in certain cases the over-all motion can be ignored while the vibratory motion is that of a free rod.

After his analysis of the vibrating rod in the free–free case, Bernoulli presents experiments on the relative frequencies and the nodal points of the higher modes. He uses glass rods that are less than 0.8 cm thick and about 30 cm long. He marks on them the nodal points predicted in table

---

[19] On 20 October 1742 Bernoulli wrote Euler that a few months earlier he had sent his paper on free rods to the Academy (Fuss [1], p. 505). In the introduction to that paper he wrote that he had submitted his first paper on the vibrating rod not long before.

[20] Both Euler and Bernoulli give the length of a second's pendulum, $g/\pi^2$ (whose period is two seconds) as 3.166 Rhenish Feet or 3166 scruples (for example, Euler [3]). We have converted by taking $g$ to be 980.592 cm per sec$^2$, its value in Geneva.

[21] Bernoulli gives 175 half cycles/second.

[22] This discrepancy was due mainly to the measurements of force and weight. When we did the experiment with a thicker rod and used the same (spring) balance extended the same amount to measure both the force and the weight, we had agreement to within a quarter tone.

[23] The compact carillon (i.e. the *glockenspiel*, forerunner of the celesta) had been recently developed. In the place of bells it used metal bars that were supported (at nodal points) but free at both ends.

(13.25). He holds a rod gently between two fingers at a nodal point and taps it at various points. On making a slight adjustment in the place of holding and being sure that he is not tapping at another node of the same mode, he produces a pure tone. He checks that the relative pitches of these tones agree with the relative frequencies predicted by (13.18). (We did this also, using a steel rod as Bernoulli recommends. We found agreement to within a quarter tone.)

Bernoulli's experiments certainly weren't crucial in deciding for or against a theory and they certainly weren't impressive in discovering new phenomena; but if one reminds one's self anew that the physics of the pendulum condition, the mathematics of fourth order differential equations with boundary conditions, and the techniques of calculation were esoteric, new, and involved, respectively, then it is hard not to follow Bernoulli in his enthusiasm as he introduces the experimental agreement with prediction:[24]

And so I will now remove my hand from the writing table and will confirm by experiments those things that are physical in our argument. So far we have derived the results by a reasoning in two parts, the one physical in which we presumed that the vibrations of a free and percussed lamina are [actually] made in those modes that our [analysis] indicates, the other, purely geometrical, I place above all doubt. But although I do not doubt the physical reasoning—in fact, it is immediately quite plausible to anyone—it will nevertheless not be possible to have it as certain before it will have been confirmed by experiments. So I hope it will not be unappreciated if I add here some experiments which I set up and observations that I made concerning diverse sounds elicited from rods in diverse ways, especially by those not sufficiently imbued with geometry, for whom experiments are the equivalent of demonstrations.

## 13.3. Superposition

In doing his experiments on a needle clamped in a "wall" Bernoulli observes superposition:[25]

... in this experiment both sounds often coexist and are perceived simultaneously, nor is it surprising since neither [mode of] oscillation obstructs the other or is an impediment. For in whatever way the lamina is curved corresponding to one [mode of] oscillation, it can still be considered as a straight line with respect to the relative displacements of another [mode of] oscillation, since the oscillations are as though infinitely small. Therefore oscillations of any mode are made in the same way, whether the lamina is free of all other oscillations or whether it executes other oscillations simultaneously. In free lamina, the oscillations of which we will soon examine, often three or four sounds are perceived simultaneously.

---

[24] Daniel Bernoulli [8], sec. 19.
[25] Daniel Bernoulli [8], sec. 8.

Thus, Bernoulli gives a convincing reason for supposing arbitrary superpositions of simple modes of oscillation to be possible motions of the rod. One could not do better without a precise definition of possible motions (i.e. dynamical equations). Of course, Bernoulli does not give here the Principle of Superposition since he doesn't assert that all small motions are superpositions of simple modes. This he asserts only later in his debate with d'Alembert and Euler on the vibrating string.

# 14. Euler (1742)

Sometime before 1739, Euler had attempted unsuccessfully to understand the oscillations of the linked compound pendulum. He solved the problem in 1740 in correspondence with Daniel Bernoulli who had suggested to him the problem of a rigid body hanging from a string.[1] In August of 1742, Euler presented the St. Petersburg Academy with a long pedagogical paper[2] on the harmonic oscillations of the linked compound pendulum and its related (limiting) cases of a rigid body hanging from a string or from a (massive, flexible) chain. This paper appeared in 1751 in the volume for the years 1741/3. In spite of the fact that some effort was required for this work, the method used is contained in Euler's paper of 1735 on the vibrating rod (Chapter 10). That is, the pendulum condition is used as usual but converted to the balancing of torques (about the points of flexure).[3]

We will discuss the main results of this paper on the linked compound pendulum in section 14.1 and the special cases in section 14.2. We will not present his pedagogical discussion, which is not significantly different from that of his 1735 paper, except to remark that in the case of one degree of freedom he deals with force explicitly from the point of view of function theory: He considers the force to be a power expansion in the displacement (whence he sees the linear force to be appropriate for small vibrations).[4]

## 14.1. Linked Compound Pendulum

Euler is concerned with three-dimensional rigid bodies; but he really treats only the two-dimensional case (or the three-dimensional case in

---

[1] On 5 November 1740 Daniel Bernoulli wrote to Euler: "I never doubted that you would solve my problem of the oscillations of bodies suspended from a flexible thread as soon as you investigated it in earnest. I am glad that this problem now seems to you to be of greater importance." (Fuss [1], II, p. 464.)

[2] Euler [8].

[3] Although moments of inertia do not arise in the vibrating rod problem, Euler did, in the 1735 paper (Chapter 10), make illustrative use of the torque balancing method to treat the rocking problem where the moment of inertia did arise in connection with the harmonic torque. In the present paper he refers explicitly to "moments of inertia."

[4] par. 6.

which there is reflective symmetry through the plane). Thus he treats the case of the uniform sphere; but in general he only hopes that the errors of the two-dimensional treatment are small. (He does not yet understand moments of inertia in three-dimensions or know about the principle axes.) We will follow his discussion only in two-dimensional notation.

Let us begin by extracting Euler's procedure for calculating torques about flexure points: Suppose that a body $\mathcal{B}$ in the $z$–$y$ plane, with mass density $\rho(z, y)$, has been displaced from an equilibrium position with its center of mass at the origin by a small rotation about $(R, 0)$ so that its center of mass is at $(0, \gamma)$, to first order in $\gamma$. Then, to first order in $\gamma$, the general point $(z, y)$ of the body has been displaced by $\gamma/R(y, -z + R)$ and the corresponding harmonic force is

$$-\mathfrak{a}^{-1}\frac{\gamma}{R}(y, -z + r)\rho(z, y)\, dz\, dy \qquad (14.1)$$

acting on the mass element $\rho(z, y)\, dz\, dy$. The harmonic torque of this force on the body, about the point

$$(S, y_0), \qquad (14.2)$$

is then

11056

$$-\mathfrak{a}^{-1}\frac{\gamma}{R}\iint_{\mathcal{B}} (z - S, y - y_0) \wedge (y, -z + R)\rho\, dz\, dy$$

11057

$$= \mathfrak{a}^{-1}\frac{\gamma}{R}\iint_{\mathcal{B}} \{z^2 + y^2\}\rho\, dz\, dy + \mathfrak{a}^{-1}M\gamma\left\{S - \frac{\gamma y_0}{R}\right\}. \qquad (14.3)$$

To first order in $\gamma$, the integral is the moment of inertia about the center of mass, $I$, and the harmonic torque is

$$\mathfrak{a}^{-1}M\gamma\left\{S + \frac{I}{RM}\right\}. \qquad (14.4)$$

This is a simple and memorable result that generalizes the more familiar case where $R = S$ that arises for the compound pendulum.

On the other hand, if there is a gravitational field $(-\mathfrak{g}, 0)$, the gravitational torque about the point $(S, y_0)$ is

$$\mathfrak{g}M\{\gamma - y_0\}. \qquad (14.5)$$

In the case of the linked compound pendulum, Euler obtains the pendulum condition simply by equating sums of harmonic and gravitational torques given by (14.4) and (14.5) as we shall now see.

Consider the linked compound pendulum of $n$ rigid bodies hanging in the $z$–$y$ plane in the gravitational field $(-\mathfrak{g}, 0)$ as follows:[5] Hang the $n$th

---

[5] pars. 46–54.

body from the point $(z_n, 0)$ and, with it in stable equilibrium so that its center of mass is at $(z_n - G_n, 0)$ with $G_n > 0$, attach the $(n-1)$st body at the point $(z_{n-1}, 0)$ with $\Delta z_n = z_n - z_{n-1} \geqslant G_n$, and continue hanging the remaining masses in this way. Thus, in its equilibrium position, the $i$th body hangs from the $(i+1)$st at $(z_i, 0)$ and has its center of mass at $z_i - G_i$ with

$$\Delta z_i = z_i - z_{i-1} \geqslant G_i > 0, \qquad (14.6)$$

where $z_0$ is chosen for convenience so that this is so. When the linked compound pendulum has undergone a small displacement, the hanging points are at $(z_i, y_i)$ respectively, with $y_n = 0$, and the centers of mass are at $(z_i - G_i, \gamma_i)$. Suppose that the straight line of the displaced $i$th body, which goes through the points $(z_{i-1}, y_{i-1})$, $(z_i - G_i, \gamma_i)$, and $(z_i, y_i)$ intersects the $z$-axis at $z_i + k_i$. This line has the tangent slope

$$\frac{y_i - \gamma_i}{G_i} = -\frac{\gamma_i}{G_i + k_i} = \frac{\Delta y_i}{\Delta z_i}. \qquad (14.7)$$

The $i$th body has been displaced from its equilibrium position by a rotation about $(z_i + k_i, 0)$. This point lies a distance $R_i = G_i + k_i$ from the center of mass which has been displaced by $\gamma_i$. Thus Euler expresses the displacement in such a way that the simple expression (14.4) is applicable: the $i$th body has the harmonic torque about the $j$th flexure point $(z_j, y_j)$

$$\mathfrak{a}^{-1} M_i \gamma_i \left\{ z_j - z_i + G_i + \frac{I_i}{(G_i + k_i) M_i} \right\} \qquad (14.8)$$

where $I_i$ is the moment of inertia of the $i$th body about its center of mass. By (14.5), the $i$th body has the gravitational torque about $(z_j, y_j)$,

$$\mathfrak{g} M_i \{ \gamma_i - y_j \}. \qquad (14.9)$$

The total torques about the $j$th flexure point are obtained by summing the torques (14.8) and (14.9) respectively over $i = 1, 2, \ldots j$. The pendulum condition is equivalent to the equality of these total torques for all flexure points[6] $(z_j, y_j)$, $j = 1, 2, \ldots n$:

$$\mathfrak{a}^{-1} \sum_{i=1}^{j} M_i \gamma_i \left\{ z_j - z_i + G_i + \frac{I_i}{(G_i + k_i) M_i} \right\} = \mathfrak{g} \sum_{i=1}^{j} M_i \{ \gamma_i - y_j \} \quad (14.10)$$

---

[6] In Euler's notation: for example, in the case $n = 3$, $j = 2$,

$$R(Gg - Aa) + Q(Ff - Aa) = \frac{R\,\mathsf{Gg}}{f} \left( AB + BG + \frac{k^2}{MG} \right) + \frac{Q\mathsf{Ff}}{f} \left( AF + \frac{i^2}{LF} \right),$$

where $R$ and $Q$ are our $M_1$ and $M_2$, respectively, Gg and Ff are our $\gamma_1$ and $\gamma_2$, Aa is our $y_2$, AB is our $z_2 - z_1$, BG and AF are our $G_1$ and $G_2$, MG and LF are our $(G_1 + k_1)$ and $(G_2 + k_2)$, $k^2$ and $i^2$ are our $I_1/M_1$ and $I_2/M_2$, and $f$ is our $\mathfrak{ag}$. (Euler explicitly calls $Rk^2$ and $Qi^2$ "moments of inertia.")

# 15. Johann Bernoulli (1742)

Throughout the era that we are studying, Johann Bernoulli held presence *the* great teacher and as a central figure in the scientific community. His correspondence includes more than 2000 letters exchanged with more [than] 100 scholars[1]). Thus, if we look for a contemporary but comprehending [re]action to the work of Euler and Daniel Bernoulli on vibrating systems, [w]e could not be more fortunate than to have Johann Bernoulli's own [re]action. This he provides. Indeed, his collected works, published in 1742, [co]ntain a volume of previously unpublished notes which include a study [o]f vibrating systems.[2] It seems that he made this study around the year [1]740.[3] One can presume that he had heard of Euler's and Daniel Bernoulli's [re]sults from before; but he probably first read their great works on the [h]anging chain at this time.[4] The vibrating systems in question include a [n]umber with a single degree of freedom that we will mention briefly, as [w]ell as the dangling rod which we will discuss in section 15.2 and the linked [p]endulum with two treatments which we will discuss in section 15.3 and [s]ection 15.4 respectively. His two treatments of the linked pendulum could [re]present his ideas on the subject from before and after his reading of [E]uler's and Daniel Bernoulli's works.

## 15.1. One Degree of Freedom

The problems in a single degree of freedom that Johann Bernoulli treats [a]re those of a body rocking on a surface without slipping, of a floating [b]ody rocking or bobbing, and of a rod that hangs from two strings such [t]hat in equilibrium the rod is horizontal and both strings are vertical.[5] The

---

[1] Spiess [1], pp. 57–58.
[2] Johann Bernoulli [1], IV, pp. 286–331.
[3] On 16 April 1740 Bernoulli wrote Euler: "what I have said [concerning the oscillations [o]f floating bodies] I have written from the manuscript which I have prepared concerning both [th]is and many other new and curious subjects pertaining to dynamics," (Eneström [1], VI, [p.] 57).
[4] Also in the letter of 16 April 1740, Bernoulli complained that he had still not received [v]olumes V and VI of the St. Petersburg *Commentarii*. Volume VI, published in 1738, contained [D]aniel Bernoulli's first paper on the hanging chain. (Eneström [1], VI, p. 59.)
[5] "De oscillationibus corporum titubantium . . . ," pp. 296–301; "De corporum aquae [in]sidentium oscillationibus . . . ," pp. 286–297; "De pendulis sympathicis," pp. 310–313.

---

which, as a set of equations, have to be taken in conjunction with (14.7) to relate the variables $\gamma_i$, $k_i$, and $y_i$. By (14.7), the equation (14.10) for $j = 1$ is

$$\mathfrak{a}g = \frac{I_1}{G_1 M_1} + G_1 + k_1. \tag{14.11}$$

## 14.2. Dangling Rod and Weighted Chain

Actually, before coming to the general set of equations (14.10), which he writes out only for $n = 1$, 2, and 3 (without summation signs or subscripts), Euler first discusses special cases of the linked compound pendulum. For example, he begins with the compound pendulum,[7] the $n = 1$ case, where there is only one equation, namely (14.11) with $k_1 = 0$. The case of the dangling rod[8] is that in which $n = 2$ and $M_2 = 0$; then Euler has, in (14.10), two equations, namely (14.11) and

$$\mathfrak{a}g = \frac{I_1}{(G_1 + k_1) M_1} + \Delta z_2 + G_1. \tag{14.12}$$

Since $I_1$ can be eliminated between equations (14.11) and (14.12), Euler has that

$$\mathfrak{a}g = \Delta z_2 \frac{G_1 + k_1}{k_1}. \tag{14.13}$$

Thus, in equations (14.11) and (14.13), Euler obtains Daniel Bernoulli's equations for the dangling rod (see equations[9] (12.28) and (12.29)). We said before that these equations have a certain elegance about them. Euler gives the following interpretation of (14.13): As the dangling rod swings, its center of mass swings as though it were a mass point swinging as a simple pendulum on an imaginary string that is always parallel to the actual string. To see this one has only to scale the linear dimensions in the plane by the factor $(G_1 + k_1)/k_1$, keeping $(z_1 + k_1, 0)$ the fixed point. See Figure 14.1.

Euler treats also the case of a rigid body hanging from a flexible chain.[10] Let the chain hang as usual from the point $(l, 0)$ in the $z$–$y$ plane so that its end is at the origin when it is in equilibrium. Let the rigid body be attached at the end so that in equilibrium its center of mass is at $(-G, 0)$. When the system is displaced, let the displaced chain be described by the function $y(z)$ and let the displacement of the center of mass of the rigid

---

[7] pars. 25–28.
[8] pars. 33–40.
[9] $I_0$ in (12.29) is $I_1 + M G_1^2$.
[10] pars. 41–45.

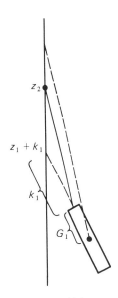

Figure 14.1

Thus, Euler obtains the total harmonic torque about $(z, y($

$$\mathfrak{a}^{-1} \int_0^z y(\hat{z})(z - \hat{z})\rho(\hat{z})\,d\hat{z} + \mathfrak{a}^{-1}M\gamma\left\{z + G + \frac{1}{(G+}\right.$$

and the total gravitational torque

$$\mathfrak{g}\int_0^z (y(\hat{z}) - y(z))\rho(\hat{z})\,d\hat{z} + \mathfrak{g}M\{\gamma - y(z)\}.$$

He has, as before, that the pendulum condition is equivaler of the expressions (14.19) and (14.20) for all $z \in [0, l]$. Eul the equality twice, obtaining the equation

$$-\mathfrak{a}^{-1}\rho(z)y(z) = \mathfrak{g}\left\{\int_0^z \rho\,dz + M\right\}y''(z) + \mathfrak{g}\rho(z)y$$

which has to be taken with the boundary conditions tha the equality of (14.19) and (14.20), namely

$$\begin{cases} \mathfrak{a}^{-1}\gamma\left\{G + \dfrac{I}{(G+k)M}\right\} = \mathfrak{g}\{\gamma - y(0)\} \\ \mathfrak{a}^{-1}\gamma = -\mathfrak{g}y'(0), \end{cases}$$

as well as, of course, $y(l) = 0$. When $M = 0$, as Euler not (14.21) is the equation of the hanging chain that he and had obtained in their papers of 1733, 1734, and 1736 (Cl

body be denoted by $\gamma$. Let $(k, 0)$ be the point at which the line of the body, through $(-G, \gamma)$ and $(0, y(0))$, crosses the $z$-axis. Then

$$\frac{\gamma - y(0)}{G} = \frac{y(0)}{k} \tag{14.14}$$

relates the variables, being the tangent of the line. Euler obtains the pendulum condition by exactly the same method that he uses for the linked pendulum. The harmonic torque of the element $\rho(\hat{z})\,d\hat{z}$ of the chain about the point $(z, y(z))$ is

$$\mathfrak{a}^{-1}y(\hat{z})(z - \hat{z})\rho(\hat{z})\,d\hat{z} \tag{14.15}$$

by (14.4), since the element has no moment of inertia about its center of mass; the corresponding gravitational torque is

$$\mathfrak{g}(y(\hat{z}) - y(z))\rho(\hat{z})\,d\hat{z} \tag{14.16}$$

by (14.5). The harmonic torque of the rigid body about the same point is

$$\mathfrak{a}^{-1}M\gamma\left\{z + G + \frac{I}{(G+k)M}\right\}, \tag{14.17}$$

by (14.4), supposing $M$ and $I$ to be its mass and moment of inertia about its center of mass; the corresponding gravitational torque is

$$\mathfrak{g}M\{\gamma - y(z)\}. \tag{14.18}$$

rod swings in its own line. What is disappointing is that he does not consider the problems in two degrees of freedom that are very naturally associated with these systems. Through Euler, at least, Johann Bernoulli knew of Daniel Bernoulli's analysis of the asymmetric floating body where rocking and bobbing are combined in the simple modes (see section 12.2). He clearly didn't take his son's analysis seriously and he dealt only with the symmetric case (in which the $r^2 - l^2$ term in (12.16) and (12.20) drops out at least when integrated in the third dimension). The transversal oscillations of the rod hanging from two strings involve two degrees of freedom, e.g. displacement of the center of mass and rotation, both being combined in the simple modes if the strings are of unequal lengths. In the problem of the rocking of a body on a surface, Bernoulli considers only the case of a two-dimensional body on a curve, though he generalizes from Euler's case (section 10.1) where the curve was required to be straight. The fact that these problems in two degrees of freedom were largely overlooked seems worth pointing out since from a modern point of view one might expect them to stand out with paramount interest in the effort to generalize dynamics from the case of one degree of freedom.

## 15.2. Dangling Rod

Johann Bernoulli's treatment of the dangling rod[6] is elegant; but it makes a great separation between the mechanics of the problem and the *momentum principle*. It is based on the notion of the spontaneous center of rotation. Before taking up Bernoulli's treatment of the dangling rod we will go back a few sections in his collected works to consider his treatment of the spontaneous center of rotation.[7]

Consider a body that lies in the $z$–$y$ plane with its center of mass at the origin. Let the body have mass $M$ and moment of inertia $I$ about the center of mass. Suppose that a force $(0, f)$ acts on the body at $(A, 0)$ and let $(0, h)$ be the force that acts at $(-B, 0)$ so as to keep the body fixed at this point. If $h$ is initially zero, then $(-B, 0)$ is called the "spontaneous center of rotation,"[8] and

$$MAB = I \qquad (15.1)$$

as Bernoulli shows as follows: The complete (including the inertial) torque about $(A, 0)$ is

$$-(A + B)h - I\ddot{\alpha} + AM\ddot{y} = 0 \qquad (15.2)$$

---

[6] "De pendulo luxato, . . . ," pp. 302–309.

[7] "De centro spontaneo rotationis," pp. 265–273.

[8] Correspondingly, $(A, 0)$ is known as the "center of percussion."

where $y$ is the displacement of the center of mass and $\alpha$ is the angle of the body's rotation. The condition that the body remain at rest at $(-B, 0)$ is, for small times,

$$\alpha = y/B. \tag{15.3}$$

With this (15.2) solved for $h$ becomes[9]

$$h = -\frac{I - MAB}{A + B} \ddot{\alpha}, \tag{15.4}$$

which yields the condition (15.1) for zero $h$. Bernoulli's treatment differs in that he saves the integrations that yield $I$ and $M$ until the end. Note that equation (14.2) because of the condition (15.3) is essentially an application of the momentum principle for one degree of freedom, which now drops out of the picture.

We pause to remark that Bernoulli takes a considerable interest in the relation (15.1). It is, as he emphasizes, also the Huygens result for the compound pendulum: If the body is hung at $(A, 0)$, its center of oscillation will be at $(-B, 0)$, and vice versa. Bernoulli considers it of great significance that he can obtain (15.1) also from a minimum principle as follows: Let $\mathfrak{M}(B) = (I + MB^2)(A + B)^{-2}$ so that a mass $\mathfrak{M}$ at $(A, 0)$ has the same moment of inertia about $(-B, 0)$ as does the original body and the force $(0, f)$ produces the same acceleration on $\mathfrak{M}$ as it does at $(A, 0)$ on the original body with a fulcrum at $(-B, 0)$. Now, $\mathfrak{M}(B)$ takes on its minimum value,

$$\mathfrak{M}_{\min} = M \frac{B}{A + B} \tag{15.5}$$

when (15.1) holds. That is, the force $(0, f)$ produces the greatest acceleration when there is no force at the fulcrum working against it. Bernoulli writes:

Behold in this an example of Nature operating in the most simple way. . . . However much final causes are commonly banned from physics, we still cannot marvel enough because the effects of Nature, clear from purely mechanical laws, conspire with the most general metaphysical canon that tells us that Nature does nothing in vain, that it always acts by the shortest route, that it squanders nothing of its force unless it is necessary for producing some effect, that Nature never uses a lot for those things that can be accomplished by a little, [etc.]."

To discuss Bernoulli's treatment of the dangling rod, we suppose that the string is fixed at $(l, 0)$ and, when the system is in equilibrium, that the string is attached to the end of the rod at the origin, with the rod's center

---

[9] In Bernoulli's notation: for a rigidly connected two-dimensional array of mass points, where a force acts at the point A, B is the center of rotation, and $P$ is "the force of impression . . . by which B is urged normally to AB," to compensate for the inertial torque about A due to $M$, $\int P = \int \mathrm{BF} \cdot M - \int \mathrm{BM}^2 \cdot M/AB$, with $\int \mathrm{BF} \cdot M = \mathrm{BC} \cdot S$, where $S$ is our $M$, BC our $B$, AB our $(A + B)$, $\int \mathrm{BM}^2 \cdot M$ our $(I + MB^2)$, and $\int P$ our $h$, with $\ddot{\alpha} = 1$.

of mass at $(-G, 0)$. Bernoulli considers the case of small isochronous oscillations so that the motion is described by scaling in the $y$-direction. Bernoulli considers the rod in its extreme position in which we will let $y$ denote the displacement of the attached end and $(k, 0)$ the intersection of the line of the rod with the $z$-axis. Since the longitudinal forces cancel, Bernoulli considers the force components that are normal to the rod. The string being at tension $\mathfrak{g}M$ exerts the normal force at the attached end $\mathfrak{g}M\{(y/k)-(y/l)\}$; and[10] gravity exerts at the center of mass the normal force $-\mathfrak{g}M(y/k)$. The center of these two forces is a distance $A$ below the center of mass where they produce no moment[11]

$$0 = \mathfrak{g}M\left\{\frac{y}{k}-\frac{y}{l}\right\}(G+A)-\mathfrak{g}M\frac{y}{k}A. \qquad (15.6)$$

Thus it is as though the rod were being restored by a normal force that acts at a point of the rod a distance

$$A = G\left\{\frac{l}{k}-1\right\} \qquad (15.7)$$

below the center of mass.[12] This force causes the rod to rotate about $(k, 0)$, that is about a point a distance

$$B = G+k \qquad (15.8)$$

above the center of mass. Thus, as Bernoulli writes, "we have the case explained . . . concerning the nature of the center of rotation. . . . The point which is to be thought of . . . as the point of suspension of an ordinary pendulum . . . will simultaneously be the spontaneous center. . . ." That is, $A$ and $B$ satisfy (15.1) where $I$ is the rod's moment of inertia about its center of mass. Thus Bernoulli obtains the following equation[13] for $k$:

$$MG\left\{\frac{l}{k}-1\right\}(G+K)=I \qquad (15.9)$$

---

[10] In Bernoulli's notation: the normal force of gravity is (BF/AB)$\mathfrak{g}M$, where BF is our $y$ and AB is our $k$; and the normal force due to the string is (AN/AB)$\mathfrak{g}M$. (A, B, and 0 are respectively the points in our notation $(k, 0)$, $(0, y)$ and $(l, 0)$; AN/AB is the angle at B which is the difference of the supplement at A, $(y/k)$ and the angle at 0, $(y/l)$.)

[11] In Bernoulli's terminology: "These two normal forces, one of which is negative with respect to the other, will have in common a center of equilibrium D below C [the center of gravity] which must be such that the normal force at C is to the normal force at B [the attached end of the rod] as BD to CD."

[12] In Bernoulli's notation: CD = ((OA · BC)/AB), where OA is our $(l-k)$, BC is our $G$, and AB is our $k$.

[13] In Bernoulli's notation:

$$\frac{ab}{x}-b=\frac{\int t^2 p}{(b+x)M},$$

where $a$ is our $l$, $b$ is our $G$, $x$ is our $k$, and $\int t^2 p$ is our $I$.

which he solves for the two solutions. To find the length $\mathfrak{a}\mathfrak{g}$ of the simple isochronous pendulum, Bernoulli simply considers that the total horizontal force $-\mathfrak{g}M(y/l)$ is acting on the mass $\mathfrak{M}_{min}$, given in (15.5), at the location on the rod a distance $A$ below the center of mass, which is displaced by $(y/k)(A+B)$. This is simple harmonic motion with the intensity

$$\mathfrak{a}^{-1} = -\frac{\left(-\mathfrak{g}M\frac{y}{l}\right)}{\mathfrak{M}_{min}\frac{y}{k}(A+B)} = \mathfrak{g}\frac{k}{l(G+k)}, \tag{15.10}$$

by (15.5) and (15.8).

## 15.3. Linked Pendulum I

Here we will discuss Bernoulli's first method for treating the linked pendulum.[14] In this method, Bernoulli attempts to say something about the motion even when it is not restricted by the assumption of small oscillations. Initially at least, the method is based on the determination of the energy exchanges between adjacent masses during the motion. In discussing this method we will modify our usual notation and let the pendulum hang from the origin in the $z$–$y$ plane, with the gravitational field acting in the positive $z$-direction, and we will number the masses from the top down. Let the $i$th link be specified by the vector

$$\mathbf{k}_i = (\sqrt{k_i^2 - r_i^2}, r_i), \qquad k_i \text{ fixed} \tag{15.11}$$

(where the square root will be positive in the case of small oscillations). The position of the $i$th mass, $m_i$, is then

$$\mathbf{X}_i = \sum_{j=1}^{i} \mathbf{k}_j. \tag{15.12}$$

Let the tension in the $i$th link be $T_i$ so that the work done by this link on the $i$th mass will be $-\int T_i \, dx_i$ where Bernoulli defines $dx_i$ by[15]

$$\frac{dx_i}{dt} = k_i^{-1}\mathbf{k}_i \cdot \frac{d\mathbf{X}_i}{dt} = k_i^{-1}\mathbf{k}_i \cdot \frac{d\mathbf{X}_{i-1}}{dt} \tag{15.13}$$

---

[14] "De pendulis multifilibus," pp. 313–324.

[15] In Bernoulli's presentation: $dx$ is the orthogonal component, in the link's own direction, of the change of position of either end of the link during time $dt$; he displays $dx$ in a geometrical figure. He uses the same notation for each of the links. (pp. 314, 319).

and the work done on the $(i-1)$st mass is $\int T_i\, dx_i$. Bernoulli has that the tangential forces on the $i$th mass due to the $i$th link and the $(i+1)$st link are respectively[16]

$$-T_i k_i^{-1} v_i^{-1} \mathbf{k}_i \cdot \frac{d\mathbf{X}_i}{dt} = -T_i v_i^{-1} \frac{dx_i}{dt} \tag{15.14}$$

and

$$T_{i+1} k_{i+1}^{-1} v_i^{-1} \mathbf{k}_{i+1} \cdot \frac{d\mathbf{X}_i}{dt} = T_{i+1} v_i^{-1} \frac{dx_{i+1}}{dt} \tag{15.15}$$

with

$$|v_i| = \left|\frac{d\mathbf{X}_i}{dt}\right| \quad \text{and} \quad \text{sign } v_i = \text{sign } \frac{dr_i}{dt}. \tag{15.16}$$

Similarly, Bernoulli has that the potential energy of the $i$th mass due to gravity is[17]

$$g m_i q_i \tag{15.17}$$

where

$$q_i = \sum_{j=1}^{i} \{k_j - \sqrt{k_j^2 - r_j^2}\} \tag{15.18}$$

and that the tangential force on the $i$th mass due to gravity is[18]

$$g m_i v_i^{-1}(1, 0) \cdot \frac{d\mathbf{X}_i}{dt} = -g m_i v_i^{-1} \frac{dq_i}{dt} \tag{15.19}$$

by (15.11), (15.12), and (15.18). From (15.14), (15.15), and (15.19) and the *momentum law*, to which Bernoulli refers as the "Dynamical Principle," he has that[19]

$$m_i v_i \frac{dv_i}{dt} = -g m_i \frac{dq_i}{dt} - T_i \frac{dx_i}{dt} + T_{i+1} \frac{dx_{i+1}}{dt}. \tag{15.20}$$

---

[16] In Bernoulli's notation: the tangential force on the top mass due to the second link is $T(dx/dr)$, where $T$ is our $T_2$, and $dr$ is our $v_1\, dt$ (since Bernoulli defines $r$ as arc length along the trajectory). Bernoulli does not give general expressions when there are more than two links.

[17] In Bernoulli's notation: $y$, $q$, $t$, etc., our $q_i$, for $i = 1, 2, 3, \ldots$, are changes of height indicated in his figure. (Bernoulli does not use the term "potential energy.")

[18] In Bernoulli's notation: the gravitational force on the top mass is $g a\,(dy/dr)$, where $a$ is our $m_1$.

[19] In Bernoulli's notation: for the top mass,

$$\frac{g a\, dy - T\, dx}{a} = v\, dv,$$

where $T\, dx$ is our $T_2\, dx_2$ and $v$ is our $v_1$ ($T_1$ never enters (15.20) because $dx_1/dt$ is zero). Bernoulli's equality is wrong by a factor of $(-1)$.

We cannot avoid the impression that Bernoulli is willing to consider the momentum law only because he succeeds in writing it in integrable form thanks to the use of quantities that are motivated from energy considerations. In the two body case which he treats first, Bernoulli writes out only one of the equations (15.20) and he immediately writes out the integrals of both.[20] In general we obtain

$$\Delta \tfrac{1}{2} m_i v_i^2 = -\Delta g m_i q_i - \int T_i \, dx_i + \int T_{i+1} \, dx_{i+1}. \tag{15.21}$$

Unless there is only one mass, the equations (15.20) or (15.21) alone do not serve as a system of dynamical equations because the $T_i$'s are not known functions. In the case of two masses, Bernoulli attempts to find $T_2$ for the general situation and we will come back to this later.

To treat in detail the case of small oscillations, Bernoulli assumes that the tensions are equal to the weights supported:

$$T_i = g \sum_{j=i}^{N} m_j, \tag{15.22}$$

if there are $N$ bodies. To second order in $r_i$,

$$q_i = \sum_{j=1}^{i} \frac{r_j^2}{2k_j} \tag{15.23}$$

and, by (15.11), (15.12), and (15.13),

$$\frac{dx_i}{dt} = k_i^{-1} \sum_{j=1}^{i-1} \left\{ -\frac{\sqrt{k_i^2 - r_i^2)}}{\sqrt{(k_j^2 - r_j^2)}} r_j \frac{dr_j}{dt} + r_i \frac{dr_j}{dt} \right\}. \tag{15.24}$$

Bernoulli assumes the equal link case,

$$k_i = k, \qquad i = 1, 2, \ldots N. \tag{15.25}$$

so that, to second order in $r_i$,

$$\frac{dx_i}{dt} = k^{-1} \sum_{j=1}^{i-1} (r_i - r_j) \frac{dr_j}{dt}. \tag{15.26}$$

On the basis of the assumption of isochronism and simultaneous crossing of the axis, Bernoulli supposes that the displacements are mutually proportional, that is that there are constants $\gamma_i$ so that[21]

$$\sum_{j=1}^{i} r_j = \gamma_i r \tag{15.27}$$

---

[20] In Bernoulli's notation: $g y - (1/a) \int T \, dx = (1/2) v^2$ and $g q + (1/b) \int T \, dx = (1/2) u^2$, where $b$ is our $m_2$ and $u$ is our $v_2$ (and where there is no indication that differences are involved; again, the equality is wrong by a factor of $(-1)$; see, however, equation (15.45)).

[21] In Bernoulli's notation: the displacements are given by $e$, $ne$, $me$, $pe$, etc., where $n$, $m$, $p$, are constants (pp. 316, 318). Thus, Bernoulli's $e$ is our $r$ if we set $\gamma_1 = 1$, and $n$, $m$, $p$, etc. are our $\gamma_i$, $i = 2, 3, 4, \ldots$

or

$$r_i = (\gamma_i - \gamma_{i-1})r, \quad \text{with } \gamma_0 = 0. \tag{15.28}$$

Whence,[22] by (15.23) and (15.25),

$$q_i = \frac{r^2}{2k} \sum_{j=1}^{i} (\gamma_j - \gamma_{j-1})^2 \tag{15.29}$$

and[23] by (15.26)

$$\frac{dx_i}{dt} = k^{-1}r\frac{dr}{dt}\left\{(\gamma_i - \gamma_{i-1})\gamma_i - \sum_{j=1}^{i}(\gamma_j - \gamma_{j-1})^2\right\}. \tag{15.30}$$

Also, by (15.27), in the small oscillation limit, since $\sum_{j=1}^{i} r_j$ is the displacement of the $i$th mass,

$$v_i = \gamma_i \frac{dr}{dt}. \tag{15.31}$$

By (15.22), (15.29), (15.30), and (15.31), the momentum law (15.20) becomes[24]

$$m_i\gamma_i^2 \frac{d^2r}{dt^2} = g\frac{r}{k}\left[-m_i \sum_{j=1}^{i}(\gamma_j - \gamma_{j-1})^2\right.$$

$$-\left(\sum_{j=i}^{N} m_j\right)\left\{(\gamma_i - \gamma_{i-1})\gamma_i - \sum_{j=1}^{i}(\gamma_j - \gamma_{j-1})^2\right\}$$

$$+\left.\left(\sum_{j=i+1}^{N} m_j\right)\left\{(\gamma_{i+1} - \gamma_i)\gamma_{i+1} - \sum_{j=1}^{i+1}(\gamma_j - \gamma_{j-1})^2\right\}\right]. \tag{15.32}$$

---

[22] In Bernoulli's notation: for example,

$$y = \frac{e^2}{2c},$$

$$q = \frac{(1+(n-1)^2)ee}{2c},$$

and

$$[t] = \frac{(1+(n-1)^2+(m-n)^2)ee}{2c},$$

where $c$ is our $k$ (pp. 315–316, 318).

[23] In Bernoulli's notation: for example, in the case of two links, $dx = ((n-2)e\,de/c)$ (pp. 316, 319).

[24] In Bernoulli's notation: for example, the *intensity* (in our notation, equal to $-\ddot{r}/r$) is, from the second mass, in the three mass system,

$$g\frac{(n-1)B+(2n-m-1)C}{ncB}$$

(pp. 321–323).

Bernoulli does express the right side of (15.20) in the form given in (15.32); but he avoids the left side of (15.32). In fact, conservation of energy is the equality

$$g \sum_{i=1}^{N} m_i q_i + \frac{1}{2} \left( \sum_{i=1}^{N} m_i \gamma_i^2 \right) \left( \frac{dr}{dt} \right)^2 = \text{const.} \tag{15.33}$$

as Bernoulli writes except that he overlooks the constant.[25] The constant doesn't matter because he differentiates (15.33) with respect to $r$. We can proceed a little more directly by differentiating with respect to $t$, using (15.29), to obtain[26]

$$\frac{d^2 r}{dt^2} = -g \frac{r}{k} \frac{\displaystyle\sum_{i=1}^{N} m_i \sum_{j=1}^{i} (\gamma_j - \gamma_{j-1})^2}{\displaystyle\sum_{j=1}^{N} m_j \gamma_j^2} \tag{15.34}$$

which allows Bernoulli to express the left side of (15.32) as a quantity that is proportional to $r$. Thus, Bernoulli's final system of equations is

$$\frac{m_i \gamma_i^2}{\displaystyle\sum_{j=1}^{N} m_j \gamma_j^2} \sum_{j=1}^{N} m_j \sum_{k=1}^{j} (\gamma_k - \gamma_{k-1})^2$$

$$= m_i \sum_{j=1}^{i} (\gamma_j - \gamma_{j-1})^2 + \left( \sum_{j=i}^{N} m_j \right) \left\{ (\gamma_i - \gamma_{i-1}) \gamma_i - \sum_{j=1}^{i} (\gamma_j - \gamma_{j-1})^2 \right\}$$

$$- \left( \sum_{j=i+1}^{N} m_j \right) \left\{ (\gamma_{i+1} - \gamma_i) \gamma_{i+1} - \sum_{j=1}^{i+1} (\gamma_j - \gamma_{j-1})^2 \right\} \tag{15.35}$$

which he actually writes out only for the case[27] $N = 3$. He notes that he could have taken instead the system of ratios of the equations (15.32).

---

[25] In Bernoulli's notation: for example, for the top mass in the case of three links "the square of the velocity of the weight $B$ is

$$[g] \frac{(Ay + Bq + Ct)nn}{A + Bnn + Cmm} ,"$$

where $A$, $B$, and $C$ are our $m_i$, $i = 1, 2$, and $3$ (p. 320). (This agrees with equation (15.33) except for the constant and factors of $(-1)$ and 2.

[26] In Bernoulli's notation: for the case given in note 25, the *intensity* (which in our notation can be written as $d\dot{r}^2 / 2r\, dr$) is

$$gn \frac{d(Ay + Bq + Ct)}{(A + Bnn + Cmm)e\, de}$$

(pp. 321–322), correcting now the factor of 2, in which the values for $y$, $q$, and $t$ from note 22 are substituted (see note 25).

[27] Bernoulli asserts that the *intensity* of note 26 equals the *intensities* of note 24.

Bernoulli notes that the length $\mathfrak{ag}$ of the simple isochronous pendulum is

$$\mathfrak{ag} = k \frac{\gamma_N}{\gamma_N - \gamma_{N-1}}. \tag{15.36}$$

This follows immediately from the observation that the lowest mass swings as though it were a simple pendulum swinging from the intersection of the axis with the line of the lowest link.

When Bernoulli treats the double pendulum, which he does first, he works with the energy equations (15.21) rather than the force equations (15.20). He carries out explicitly the work integrals with the help of (15.22) and (15.30):[28]

$$\int T_i \, dx_i = \frac{\mathfrak{g}}{2k} \left( \sum_{j=i}^{N} m_j \right) \left\{ (\gamma_i - \gamma_{i-1})\gamma_i - \sum_{j=1}^{i} (\gamma_j - \gamma_{j-1})^2 \right\} \Delta r^2. \tag{15.37}$$

Similarly, from (15.29), he has the change in potential energy:

$$\Delta \mathfrak{g} m_i q_i = \frac{\mathfrak{g}}{2k} m_i \left( \sum_{j=1}^{i} (\gamma_j - \gamma_{j-1})^2 \right) \Delta r^2. \tag{15.38}$$

Let us consider the energy equations (15.21) with the limits of integration being the axial position $r = 0$ and the extreme position $r = r$ (where $dr/dt = 0$). Then, by (15.37) and (15.38), (15.21) can be written

$$-\tfrac{1}{2} m_i \gamma_i^2 \left( \frac{dr}{dt} \right]_{r=0} \right)^2 = -\frac{\mathfrak{g}}{2k} m_i r^2 \sum_{j=1}^{i} (\gamma_j - \gamma_{j-1})^2$$

$$-\frac{\mathfrak{g}}{2k} \left( \sum_{j=i}^{N} m_j \right) r^2 \left\{ (\gamma_i - \gamma_{i-1})\gamma_i - \sum_{j=1}^{i} (\gamma_j - \gamma_{j-1})^2 \right\}$$

$$+\frac{\mathfrak{g}}{2k} \left( \sum_{j=i+1}^{N} m_j \right) r^2 \left\{ (\gamma_{i+1} - \gamma_i)\gamma_{i+1} - \sum_{j=1}^{i+1} (\gamma_j - \gamma_{j-1})^2 \right\}$$

$$\tag{15.39}$$

and Bernoulli considers the ratios of these equations. In the $N = 2$ case that he looks at, this is then the ratio

$$\frac{m_1}{m_2 \gamma^2} = \frac{-m_1 + m_2(\gamma - 2)}{-m_2(\gamma^2 - \gamma)} \tag{15.40}$$

if $\gamma_1 = 1$ and $\gamma_2 = \gamma$, which has the solutions $\gamma = 1 \pm \sqrt{(m_1 + m_2)/m_2}$, as Bernoulli finds.

Our derivation of equations (15.39) is an idealization of what Bernoulli does. Since Bernoulli attempts to carry out his derivation in part for the case of non-small oscillations, as we mentioned above, we shall now go

---

[28] That is, for example, $\int dx_2 = (1/2k)(\gamma_2 - 2)\Delta r^2$. In Bernoulli's notation: $((n-2)ee/2c) = \int dx$ (p. 317).

back and follow him more closely, in his case $N = 2$ only. Equations (15.21) are then

$$\Delta \tfrac{1}{2} m_1 v_1^2 = -\Delta g m_1 q_1 + \int T_2 \, dx_2$$

$$\Delta \tfrac{1}{2} m_2 v_2^2 = -\Delta g m_2 q_2 - \int T_2 \, dx_2 \qquad (15.41)$$

but Bernoulli neglects to write the differences, denoted here by $\Delta$. Neglecting to notate this is a trivial matter if the two terms are lumped together as $\Delta \{ \tfrac{1}{2} m_1 v_1^2 + g m_1 q_1 \}$; but it seems to be more than notational in the case at hand. Bernoulli takes the ratio to obtain

$$\int T_2 \, dx_2 = g m_1 m_2 \frac{\Delta v_2^2 q_1 - \Delta v_1^2 q_2}{m_1 \Delta v_1^2 + m_2 \Delta v_2^2} \qquad (15.42)$$

(without the $\Delta$'s). He solves for $T_2$ by differentiating; but as soon as he substitutes for $T_2$ he integrates again and he can as well keep the form (15.42). Now one can obtain a second equation for $T_2$ from the momentum law by looking at the normal components:

$$v_2^{-1} \frac{d\mathbf{X}_2}{dt} \wedge \left\{ -T_2 k^{-1} \mathbf{k}_2 + (g m_2, 0) - m_2 \frac{d^2 \mathbf{X}_2}{dt^2} \right\} = 0 \qquad (15.43)$$

or, as is clear from (15.14) and (15.19),

$$-T_2 v_2^{-1} \sqrt{ \left( v_2^2 - \left( \frac{dx_2}{dt} \right)^2 \right)} + g m_2 v_2^{-1} \sqrt{ \left( v_2^2 - \left( \frac{dq_2}{dt} \right)^2 \right)} - m_2 v_2^2 \kappa_2 = 0, \quad (15.44)$$

with $\kappa_2$ the curvature of the path taken by the mass $m_2$ (where signs are not specified). Explicitly, Bernoulli sets out to find $T_2$ by looking at the normal components of the forces; but this time he doesn't mention the momentum law or any "Dynamical Principle." He obtains equation (15.44) but without the last term[29] which is the centrifugal force. This is only an oversight since he later makes a point of the fact that the centrifugal forces can be neglected in the small oscillation limit. Bernoulli solves the equation (15.44) for $T_2$ and substitutes this into equation (15.42). With the corrections that we have included, this gives an equation that is correct for general motions of the double pendulum. Had Bernoulli substituted $T_2$ as found from equation (15.44) into the equations (15.41) instead of (15.42) and had he made explicit the connection with the configuration (e.g. as given in (15.11) and (15.12)), Bernoulli would have provided a complete set of dynamical equations for the double pendulum on the basis of the momentum principle. It seems to us that his failure to do this illustrates quite well

---

[29] In Bernoulli's notation:

$$T \frac{\sqrt{(ds^2 - dx^2)}}{ds} = \frac{g b \sqrt{(ds^2 - dq^2)}}{ds}$$

(p. 315).

the contemporary perspective on the momentum law: In fact, Bernoulli seems interested in the momentum law (15.20) only because he is writing (half of) it in an immediately integrable form—not because he sees it as a fundamental or generally useful starting point. It was already common practice to use the momentum principle, in the case of one degree of freedom, when it is in integrable form (cf. Hermann's work, Chapter 5). It was also common practice to use the momentum law, as a restricting condition, in the case of several degrees of freedom. What Bernoulli does that is new is to put these two things together; but he does so more as a matter of manipulation than as a matter of principle.[30] He eliminates $T_2$ by chance manipulations without expecting it to be a possibility on principle. We should also mention that Bernoulli later finds the standard method of the pendulum condition to be "the most natural [method] of all" for treating the small oscillations, as we shall see in section 15.4.

For the case of small oscillations, equation (15.44), with or without the centrifugal term, gives $T_2 = \mathfrak{g}m_2$. With this substituted in (15.42) Bernoulli has a correct equation, corresponding to the ratio of equations (15.39) when the other consequences of the small oscillation assumption are introduced—except that Bernoulli writes (15.42) without the differences, writing $v_i^2$ for $\Delta v_i^2$ and $q_i$ for $\Delta q_i$. He is able to overlook this error for the following reason: If the integration is taken from a state of zero potential energy to a state of zero kinetic energy, then (15.42) takes the form

$$\mathfrak{g}m_2 \int dx_2 = \mathfrak{g}m_1 m_2 \frac{\hat{v}_2^2 q_1 - \hat{v}_1^2 q_2}{m_1 \hat{v}_1^2 + m_2 \hat{v}_2^2} \tag{15.45}$$

where $\hat{v}_i$ denotes the initial velocities and $q_i$, the final heights. With the substitutions from (15.29), (15.31), and (15.37), which Bernoulli makes, (15.45) is then the ratio of the equations (15.39) in the $N = 2$ case.

## 15.4. Linked Pendulum II

We come now to Bernoulli's second[31] and third[32] methods for obtaining the equations for the relative displacements of the linked pendulum. The first of these is ostensibly based on the analogy between the isochronous oscillations of the pendulum in two dimensions with the stable rotations of the pendulum about its axis, which means simply that the negative of the harmonic force is replaced by centrifugal force. The second is explicitly based on the pendulum condition. Since the methods that we have to discuss here are by now rather old, we will be brief.

---

[30] For a less restrained opinion see Truesdell [1], who considers this to be "a great advance in principle" (p. 184).

[31] "De pendulis multifilibus," pp. 325–329.

[32] ibid., 329–331.

We return to the notation that we used in discussing Euler and Daniel Bernoulli's treatment: Let the pendulum hang in the $z$–$y$ plane from the point $(l, 0)$ with the gravitational field acting in the negative $z$-direction. Let the masses, $m_0, m_1, \ldots m_{n-1}$ be numbered from the bottom up. Let their displacements, assumed small, be denoted $y_0, y_1, \ldots y_{n-1}$, with $y_n = 0$. Let their equilibrium positions on the $z$-axis be $z_0 = 0$, $z_1, \ldots z_{n-1}$, with $z_n = l$. Write $\Delta z_{i-1} = z_i - z_{i-1}$ for the length of the $i$th link up. The $i$th link makes the angle $\alpha_i = (y_i - y_{i-1})/\Delta z_{i-1}$ with the $z$-axis. Since the displacements are assumed small, the $i$th link has the tension $g \sum_{j=0}^{i-1} m_j$ and it exerts the horizontal forces $\pm \alpha_i g \sum_{j=0}^{i-1} m_j$ on the masses $m_{i-1}$ and $m_i$ respectively. According to the pendulum condition, these forces are balanced by the negatives of the harmonic forces, $\mathfrak{a}^{-1} m_k y_k$; but in the interpretation in which the pendulum is in rotation about its axis, $\mathfrak{a}^{-1}$ is the square of the angular velocity and $\mathfrak{a}^{-1} m_k y_k$ is the centrifugal force.

When Bernoulli thinks of the problem from the rotational point of view, he balances the horizontal force of each link with the total centrifugal force of all the masses that hang from it,

$$g \left( \sum_{j=0}^{i} m_j \right) \frac{y_{i+1} - y_i}{\Delta z_i} + \mathfrak{a}^{-1} \sum_{j=0}^{i} m_j y_j = 0 \qquad (15.46)$$

which has the continuum limit, for the hanging chain

$$g \left( \int_0^z \rho \, dz \right) y'(z) + \mathfrak{a}^{-1} \int_0^z \rho y \, dz = 0, \qquad (15.47)$$

as Bernoulli says. What he actually does for (15.46) differs in that he solves first for $\mathfrak{a}^{-1}$ in the $i = 0$ equation and writes the higher equations with this solution in the place of $\mathfrak{a}^{-1}$.

On the other hand, when Bernoulli thinks of the problem from the explicit point of view of the pendulum condition, he balances forces at each mass:

$$\mathfrak{a}g \left( \sum_{j=0}^{i} m_j \right) \frac{y_{i+1} - y_i}{\Delta z_i} - \mathfrak{a}g \left( \sum_{j=0}^{i-1} m_j \right) \frac{y_i - y_{i-1}}{\Delta z_{i-1}} + m_i y_i = 0. \qquad (15.48)$$

Bernoulli gives the continuum limit of this equation as the derivative of (15.47). The equation (15.48) is, of course, the equation found by Daniel Bernoulli and Euler, (9.3).

It seems quite possible that Bernoulli appended these methods to his earlier method when he saw the papers of Daniel Bernoulli and Euler on the subject. He certainly came upon them later. The equations (15.48) and (15.46) with their derivations are starkly simple in comparison with the equations (15.35) and their derivation and Bernoulli calls the method of the *pendulum condition* the "most natural of all." No doubt, this contrast would have discouraged the thought of pursuing the *momentum principle* as a starting point for dynamics!

# Appendix: Daniel Bernoulli's Papers on the Hanging Chain and the Linked Pendulum *

* Daniel Bernoulli [4, 5]. The original publications are reproduced on pp. 125–155. Our translations are given on pp. 156–176. In the translations we have corrected some minor errors, indicated by [ ]. We have not attempted to correct all such errors. Figures appear at the end of each paper.

# COMMENTARII
## ACADEMIAE
### SCIENTIARVM
# IMPERIALIS
## *PETROPOLITANAE.*

---

### TOMVS VI.
### AD ANNOS cɪɔ ɪɔ cc xxxɪɪ. & cɪɔ ɪɔ cc xxxɪɪɪ.

### *PETROPOLI,*
### TYPIS ACADEMIAE. cɪɔ ɪɔ cc xxxɪ:

## *Danielis Bernoulli*
# THEOREMATA DE OSCILLA-
## TIONIBVS CORPORVM FILO FLEXILI CON-
## NEXORVM ET CATENAE VERTICA-
## LITER SVSPENSAE.

## Introductio ad Argumentum.

Theoriae oscillationum, quas adhuc Auctores pro corporibus dederunt folidis, inuariatum partium fitum in illis ponunt, ita ut fingula communi motu angulari ferantur. Corpora autem, quae ex filo flexili fuspenduntur, aliam poftulant theoriam, nec fufficere ad id negotium videntur principia communiter in mechanica adhiberi folita, incerto nempe fitu, quem corpora inter fe habeant, eodemque continue variabili. De his cogitandi anfam mihi aliquando dedit catena verticaliter fuspenfa et motibus oscillatoriis agitata, hancque tunc videns motibus valde irregularibus iactari, primo mentem fubiit, ad quamnam curuam catena effet inflectenda, vt omnibus eius partibus fimul moueri incipientibus hae quoque vna in fitum peruenireut lineae verticalis per punctum fuspenfionis transeuntis : hoc modo oscillationes aequabiles fore intellexi atque tales quarum tempora definiri poffent: Mox vero fenfi difficile effe hanc determinare curuam, nifi difquifitionis initium fiat a cafibus fimpliciffimis. Q fus itaque tum has meditationes a corporibus duobus filo flexili in data diftantia cohaerentibus ; poftea tria

confi-

## DE OSCILLATIONIBUS CORPORUM. 109

conſideraui moxque quatuor, et tandem numerum eorum diſtantiasque qualescunque; cumque numerum corporum infinitum facerem, vidi demum naturam oscillan-
tis catenae ſiue aequalis ſiue inaequalis craſſitiei ſed vbique
perfecte flexilis. Suo ſingula percurram ordine; demon-
ſtrationes autem quas nunc adornare non vacat in a'i-
am occaſionum reſeruabo. In ſolutione nouis vſus ſum
principiis, proptereaque volui theoremata experimentis confirmare, ne de eorum veritate dubium eſſe poſ-
ſet, iis praeſertim, qui hisce rebus ſua natura aliquan-
to difficilioribus non omnem dare poterunt animi at-
tentionem, quique ſic facile in falſam incidere poſſent
ſolutionem. Caeterum alias oſcillationes non conſide-
rabimus, quam quae minimae ſint et iſochronae: pro experimentis tamen ſine notabili errore paulo maiores il-
las efficere licebit.

## Theorema I.

2. *Fuerit filum perfecte flexile non graue* AHF *ſuſpen-*   Fig. 1. 2.
*ſum ex puncto* A *habeatque in* H *et* F *duo alligata pon-*
*dera aequalia: tantum autem diſtet corpus inferius a ſu-*
*periori quantum hoc a puncto ſuſpenſionis, Sit porro li-*
*nea* ABC *verticalis, et ab hac corpora* H *et* F *veluti*
*infinite parum diſtent; Denique ducantur horizontales mi-*
*nimae* HB *et* FC: *Dico ſi ambo corpora ſimul oſcillari*
*incipiant, fore vt eodem temporis puncto perueniant in ſi-*
*tum lineae verticalis atque hoc modo oſcillationes ſuas*
*vniformiter perficiant, cum ſumitur* CF : BH $= 1 + \sqrt{2} : 1$.

## Corollarium.

3. Igitur duobus modis oſcillationes fiunt vnifor-
          O 3                                          mes;

mes; nempe cum fumitur, vt figura prima oftendit, $CF = (1 + \sqrt{2})BH$; tum etiam cum ad normam figurae fecundae fit $CF = (1 - \sqrt{2})BH$.

# Theorema 2.

4. *Factis ofcillationibus corporum* H *et* F *vniformibus, erit longitudo penduli fimplicis tautochroni* $= \frac{1}{2 \pm \sqrt{2}}$. AH *vel* $\frac{1}{4 \pm \sqrt{8}}$. AC, *vbi fignum affirmatiuum valet pro ofcillationibus contrariis figurae fecundae, fignum negatiuum pro confpirantibus figurae primae.*

# Corollarium.

5. Multo itaque celerius ofcillationes contrariae abfoluuntur, quam confpirantes: illarum enim 231 numerabis, dum hae centies fuerint replicatae. Confpirantes autem parum differunt ab iis quae fierent fub iisdem circumftantiis pofito filo AHF rigido: paullo tamen celerius ofcillantur corpora in filo rigido quam flexili, erunt nempe numeri ofcillationum aequali tempore peractarum praeterpropter vt 1012 ad 1000.

# Scholium.

6. Vt ad experientiam reuocarem hasce propofitiones, vfus fum globis plumbeis perfecte aequalibus, qui dum funderentur in medio tenui foramine perforati manebant; traiecto filo fericeo globisque ope nodorum firmatis ita vt inferior duplo magis diftaret a puncto fuspenfionis quam fuperior. Digitis deduxi globum inferiorem in fitum F tenfo filo: mox ofcillationes fiebant vniformes, et ope diuifionum in pariete factarum diftincte cognoui excurfiones corporum H et

F in

## DE OSCILLATIONIBVS CORPORVM.   III

F in figura 1. fuiffe vt 100 ad 241 , id eft, vt 1 ad
$1 + \sqrt{2}$ (§. 2.): Numerus etiam ofcillationum dato
tempori conueniens accurate refpondit longitudini pen-
duli fimplicis ifochroni $\frac{1}{2-\sqrt{2}}$ A H in *prop.* 4. definitae.
Deinde facta $FC = (1-\sqrt{2})$ BH in figura fecunda, de-
tinui manibus globos in fitu F et H illosque mox eodem temporis puncto dimifi: ofcillationes ortae funt
fic fatis vniformes fecus atque fiebat cum alia propor-
tione diftantiae FC et HB fumerentur: numerus ofcillationum accurate rurfus fuit, qui conueniret longitudini penduli fimplicis $\frac{1}{2+\sqrt{2}}$ A H ifochroni *prop.* 4.

## Theorema 3.

### GENERALE PRO DVOBVS CORPORIBVS.

7. *Fuerit iam pars fili* $AH = l$; $HF = L$; *pon-*
*dus corporis* $H = m$, *alteriusque* $F = M$ : *dico fore ofcil-*
*lationes vniformes fi fit*

$$CF = \frac{mL - ml + ML + Ml \pm \sqrt{(4\,mMLL + (ml + ML + Ml - mL)^2)}}{2Ml} \times BH.$$

*longitudinem autem penduli fimplicis ifochroni fore*

$$= \frac{2\,mLl}{mL + ml + Ml + ML \mp \sqrt{(4\,mMLL + (ml + ML + Ml - mL)^2)}};$$

*aut, pofito* $L + l = \lambda$ *et* $M + m = \mu$,

$$= \frac{2\,m\lambda l - 2\,m ll}{\mu\lambda \mp \sqrt{(\mu\mu\lambda\lambda - 4\,m\mu l\lambda + 4\,m\mu ll)}}.$$

## Theorema 4.

8. *Si loco duorum corporum aequalium ponantur*
*tria tantum a fe inuicem diftantia, quantum fupremum*
*a puncto fufpenfionis* A *diftat, poterunt tribus diuerfis* Fig. 3. 4. 5.
*modis ofcillationes fieri vniformes: primus eft quem figura*
*tertia*

*tertia indicat, cum pofita* $BH = 1$, *fumitur* $CF = 2$, 292 *et* $DG = 3,922$: *fecundus, qui figura quarta repraefentatur, obtinetur faciendo* $CF = 1,353$ *et* $DG = -1,044$ *ac tertius cum fit, vt in figura quinta*, $CF = -0,645$ *et* $DG = 0,122$.

Eft nempe $CF$ aequalis accipienda tribus radicibus huius aequationis

$$4x^3 - 12xx + 3x + 8 = 0,$$

tumque pro quauis radice fumenda eft

$$DG = 2xx - 2x - 2.$$

# Theorema 5.

9. *Factis, vt modo dictum, ofcillationibus vniformibus; erit longitudo penduli fimplicis ifochroni in cafu figurae tertiae proxime aequalis* $2,406\ AH$; *in cafu figurae quartae* $= 0,436\ AH$ *et in cafu figurae quintae* $= 0,159\ AH$: *Eft fcilicet longitudo penduli ifochroni* $= \frac{1}{5 - 2xx} \times AH$, *pofito rurfus* $4x^3 - 12xx + 3x + 8 = 0$.

# Scholium.

10. Ambo haec theoremata experimento accurate confirmaui in cafu figurae tertiae, deducto tantum corpore intimo extra fitum lineae verticalis mox dimittendo; quamuis enim in primis ofcillationibus inaequalitas quaedam fentiri potuerit, tamen haec fua fponte et citiffime abiit, ita vt excurfiones fingulorum corporum pluribus vicibus fucceffiuis, quantum oculis difcerni poterat, eaedem manerent: fumtis autem earundem menfuris, talis inter eas reperta fuit proportio

qua-

## DE OSCILLATIONIBVS CORPORVM. 113

qualem theorema 4. indicat: numerus quoque ofcillationum perfecte refpondit theoremati quinto: duo re-
liqui cafus maiorem induftriam requirunt: potui tamen
vtriusque generis ofcillationes fatis exacte efficere, vt
veritas theorematis quinti appareret.

# Theorema 6.
## GENERALE PRO TRIBVS CORPORIBVS.

11. *Fuerint nunc rurſus pondera corporum* H, F, G
*qualiacunque ſimulque diſtantiae eorundem a puncto ſuspen-*
*ſionis rationem habuerint qualemcunque: ſit nempe pondus*
*corporis* H $=m$, *corporis* F $=$ M, *corporisque* G $=\mu$;
*tumque* AH $= l$, HF $=$ L *et* FG $=\lambda$; *dico oſcillationes*
*vniformes futuras eſſe ſi poſita* BH $= 1$, GF $=x$ *fiat*

$((\mathrm{MM}l\lambda + \mathrm{M}\mu l\lambda)xx + (m\mathrm{M}l\lambda + m\mu l\mathrm{L} - m\mathrm{MML}\lambda$
$- \mathrm{MM}l\lambda - \mathrm{MML}\lambda + m\mu l\lambda - \mathrm{M}\mu l\lambda - \mathrm{M}\mu\mathrm{L}\lambda)x$
$- m\mu l\lambda - m\mathrm{M}l\lambda)\times((\mathrm{M}l\lambda + \mu l\lambda)x - m\mathrm{L}\lambda - \mathrm{M}l\lambda$
$- \mathrm{ML}\lambda - \mu l\lambda - \mu\mathrm{L}\lambda + ml\mathrm{L}) = mm\mu l l\mathrm{LL}x.$

*ſimulque ſumatur pro quauis radice*

$\mathrm{DG} = (\frac{\mathrm{MM}\lambda}{m\mu\mathrm{L}} + \frac{\mathrm{M}\lambda}{m\mathrm{L}})xx + (1 + \frac{\lambda}{\mathrm{L}} + \frac{\mathrm{M}\lambda}{\mu\mathrm{L}} - \frac{\mathrm{M}\lambda}{\mu l} - \frac{\mathrm{MM}\lambda}{m\mu\mathrm{L}} - \frac{\mathrm{MM}\lambda}{m\mu l} - \frac{\mathrm{M}\lambda}{m\mathrm{L}} - \frac{\mathrm{M}\lambda}{m l})x - \frac{\mathrm{M}\lambda}{\mu\mathrm{L}} - \frac{\lambda}{\mathrm{L}}.$

# Corollaria.

12. I. Ponatur maſſa corporis infimi $\mu = o$, di-
uidaturque aequatio fundamentalis fuperioris paragraphi
per factorem alterum ceu radicem inutilem; factor igitur prior erit $= o$, hincque habebitur M $lxx + (ml - $
$m\mathrm{L} - \mathrm{M}l - \mathrm{ML})x - ml = o$, vel

$x = \frac{m\mathrm{L} - ml + \mathrm{ML} + \mathrm{M}l \pm \sqrt{(4\,m\mathrm{M}ll + (ml - m\mathrm{L} - \mathrm{M}l - \mathrm{ML})^2)}}{2\,\mathrm{M}l}$;

*Tom. VI.*        P        No-

**114**                *THEOREMATA*

Notandum autem eft, non differre hunc valorem ab
illo quem dedimus in theoremate tertio, quamuis quan-
titates ab vtraque parte figno radicali inuolutae diuer-
fam habeant formam.

II. Si vero maffa corporis medii indicata per M
ponatur $= o$, tunc, vt appareat confenfus inter theorema tertium et fextum, erit in hoc pofteriori intel-
ligendum per $L + \lambda$ et $\mu$, quod defignatum fuit in
altero per L et M, ipfaque linea DG in praecedente
paragrapho definita comparanda erit cum linea CF ad
theorema tertium pertinente. Ad haec qui animum
aduerterit, vtriusque theorematis aequationes easdem
effe. reperiet inftituto calculo.

III. Denique cum ponitur corpus fummum H in-
dicatum per $m = o$, poteft in aequatione fundamentali
§. 11. vterque factor poni $= o$, et vtroque modo obtinetur CF feu $x = 1 + \frac{L}{l}$, prouti natura rei poftulat,
quia tunc lineae AH et HF, vt patet, debent in di-
rectum iacere. Excurfio autem corporis infimi ex aequatione dignofci non poteft, nifi id particulari methodo fiat. Ita nec ofcillationes definiri immediate poffunt per theorema fextum, eum duo corpora vniuntur
euanefcente alterutra longitudinum L vel $\lambda$.

IV. Fieri poteft in figura quarta, vt fit $CF = o$,
quo in cafu, quia durante tota ofcillatione diftantiae
corporum a linea verticali eandem perpetuo inter fe
rationem feruant, corpus medium F quiefcit, dum ambo reliqua hinc inde agitantur; atque. tunc perfpicuum
eft, longitudinem penduli ifochroni fore $= \lambda$, quia corpus infimum veluti ex puncto fixo C fufpenfum ofcillatur;

## DE OSCILLATIONIBVS CORPORVM. 115

latur; Iſte vero caſus, de quo loquimur, obtinetur ponendo $x = 0$, ſeu $x = \frac{mlL}{mL + Ml + ML + \mu l + \mu L}$.

# Theorema 7.

## QVOD GENERALITER PENDVLVM TAVTOCHRONVM PRO TRIBVS GORPORIBVS OSCILLANTIBVS DEFINIT.

**13.** *Retentis denominationibus et aequationibus theorematis ſexti dico oſcillationes ſingulorum corporum iſochronas fore cum oſcillationibus penduli ſimplicis, cuius longitudo ſit* $\frac{mlL}{mL + (M + \mu) \times (l + L - lx)}$.

# Corollarium.

**14.** In caſu $x = 0$, quem modo allegauimus, fit longitudo penduli iſochroni $= \frac{mlL}{mL + Ml + ML + \mu l + \mu L} = \lambda$, quod conuenit cum *Coroll.* 4. *Theorem.* 6. Si praeterea ponatur $L = l$, fit longitudo penduli iſochroni ſeu $\lambda = \frac{ml}{m + 2M + 2\mu}$: Conuenit hoc cum problemate 1. quod *Pater meus* in *Comment. Acad. Petrop. Tom. III p.* 15. dedit: idque vnicuique manifeſtum erit, qui conſiderabit pondus P, quod ibi ab vna parte chordae eſt appenſum, hic eſſe ſummam ponderum G et F auctam dimidio pondere H.

# Scholium Generale.

**15.** Poſſum ſimiles aequationes dare pro quatuor, quinque et quot libuerit corporibus: ſemper autem aequatio ad tot aſſurgit dimenſiones quot ſunt corpora,

**116** *THEOREMATA*

et eft plerumque admodum prolixa: attamen quia ac-
quatio finalis oritur ex pluribus aequationibus radicali-
bus linearibus, lex apparet ex methodo qua vfus fum,
cuius auxilio ex tempore omnia determinari poffunt,
quae ad aequationem determinandam concurrunt.

# Theorema 8.

## DE FIGVRA CATENAE VNIFORMITER OSCILLANTIS.

Fig. 6.

16. *Sit catena* AC *vniformiter grauis et perfecte
flexilis fufpenfa de puncto* A, *eaque ofcillationes facere
vniformes intelligatur: peruenerit catena in fitum* AMF;
*fueritque longitudo catenae* $= l$: *longitudo cuiuscunque partis* FM$= x$, *fumaturque* n *eius valoris, vt fit*

$$1 - \frac{l}{n} + \frac{ll}{4\,nn} - \frac{l^3}{4.9.n^3} + \frac{l^4}{4.9.16.n^4} - \frac{l^5}{4.9.16.25.n^5} + \text{etc.} = 0:$$

*Ponatur porro diftantia extremi puncti* F *ab linea verticali* $= 1$, *dico fore diftantiam puncti vbicunque affumti* M
*ab eadem linea verticali aequalem*

$$1 - \frac{x}{n} + \frac{xx}{4\,nn} - \frac{x^3}{4.9.n^3} + \frac{x^4}{4.9.16\,n^4} - \frac{x^5}{4.9.16.25\,n^5} + \text{etc.}$$

# Scholium.

17. Per methodum, quam dedi in *Comm. Acad.
Petrop. Tom. V. de refolutione aequationum fine fine progredientium*, inuenitur breuiffimo calculo $n =$ proxime o,
691 *l*: Igitur fi fuerit v. gr. punctum M in medio catenae, illud diftabit a linea verticali praeterpropter dua-
bus quintis, vel accuratius, trecentis nonaginta octo
partibus millefimis diftantiae puncti infimi F ab eadem
linea verticali. Habet autem littera *n* infinitos valores
alios.                                                    The

## DE OSCILLATIONIBVS CORPORVM. 117

# Theorema 9.

18. *Seruatis pofitionibus theorematis octaui, dico longitudinem penduli fimplicis ifochroni cum ofcillante catena effe =n, feu fubtangenti* CP *curuae* AF *in infimo puncto* F; *aut proxime aequalem fexcentis nonaginta et vni partibus millefimis totius catenae in cafu figurae fextae.*

## Corollarium.

19. Tardius igitur hoc modo ofcillatur catena, quam baculus rigidus aequabilis craffitiei, eiusdem cum catena flexili longitudinis ε huius enim ofcillationes ifochronae funt cum ofcillationibus penduli fimplicis, quod in longitudine duos baculi trientes habet.

## Scholium I.

20. Poftquam plurimos globulos plumbeos aequales ad diftantias minimas aequales filo connexi, vt in paragrapho fexto dictum, eo catenae loco vfus fum ad experimentum inftituendum: filum itaque globis oneratum ex puncto firmo fufpendi: deductaque ad latus extremitate F, eaque rurfus dimiffa rationem obferuaui, ofcillationibus iam vniformibus factis, inter diftantias puncti extremi F et medii M a linea verticali A C, eamque rationem eandem deprehendi, quae paragrapho decimo feptimo indicatur: numerum quoque ofcillationum conuenire obferuaui cum longitudine penduli fimplicis ifochroni, quae in theoremate nono exhibetur.

P 3

Scho-

**118**      *THEOREMATA*

# Scholium 2.

21. Quia aequatio in theoremate octauo exhibi-
ta, nempe

$$1 - \frac{l}{n} + \frac{ll}{4\,nn} - \frac{l^3}{4.9.\,n^3} + \frac{l^4}{4.9.16\,n^4} - \frac{l^5}{4.9.16.25.\,n^5} + \text{etc.} = 0.$$

habet infinitas radices reales, ideoque catena infinitis
modis inflecti poteft, vt ofcillationes fiant vniformes:
femper autem littera *n* minorem atque minorem valorem affumit, ita vt tandem pene euanefcat, eftque
longitudo penduli fimplicis ifochroni conftanter $= n$,
feu fubtangenti C P: vnde etiam ofcillationes tandem
fient veluti infinite celeres. Cafus qui fingi poffunt
omnes huc redeunt, primo, vt catena lineam verticalem in alio puncto non interfecet praeter punctum fufpenfionis, qui repraefentatur figura fexta et pro quo
conuenit longitudo penduli fimplicis ifochroni $n = 0$,
691 *l*, vt vidimus in antecedentibus: vel vt catena lineam verticalem in vno infuper puncto immobili fe-
cet, qualem figura feptima indicat, vbi praedictum interfectionis punctum eft B: in hoc cafu eft longitudo
penduli ifochroni $n = 0$, 13 *l*, et ofcillationes numero
viginti tres fient, dum in cafu figurae fextae decem abfoluuntur: linea C B erit proxime $= 0$, 19 *l*: C N puncto maximae excurfionis M conueniens $= 0$, 47 *l*:
ipfaque M N praeterpropter $= \frac{2}{5}$ F C. Poft hunc cafum fequitur ille, qui figura octaua fiftitur: vbi linea
verticalis in duobus punctis fixis B ct G a catena ofcillante interfecatur: deinde cum tres fiant interfectiones et fic porro. Arcus inter duo interfectionis pun-
cta proxima incepti eo maiores funt, quo altius po-
fiti:

## DE OSCILLATIONIBVS CORPORVM. 119

fiti : In catena autem infinite quafi longa arcus
fummus non differt fenfibiliter a figura chordae mu-
ficae tenfae, quia pondus iftius arcus veluti nullum
eft refpectu ponderis catenae totius. Neque difficile effe theoriam chordarum muficarum ex theoria
ifta deducere, quae plane conuenit cum illis, quas
*Taylorus* et *Pater meus* dederunt, primus in *tractat.*
*fuo de methodo incrementorum*, alter in *Comment. Acad.*
*Sc. Petrop. Tom. III.* Similes quoque interfectiones in
chordis muficis, quae in catenis vibratis effici poffe
experimentum docet, quod chartula chordae quibusdam
in locis impofita non decidat, cum chorda annexa vi-
bratur.

# Theorema. 10.

**22.** *Si catena* AC *e filo* LA *non graui fufpenfa*    Fig.. 9
*fuerit, ponaturque longitudo partis ad libitum affumtae*
F M $= x$: *diftantia fupremi puncti* N *a linea verticali*
$= \varepsilon$: *ficque porro n fumatur eius valoris vt fit*

$$1 - \frac{(l+\lambda)}{n} + \frac{(ll+\gamma l\lambda)}{4\,nn} - \frac{(l^3 + 3ll\lambda)}{4.9.n^3} + \frac{(l^4 + 4l^3\lambda)}{4.9.16\,n^4} - \text{etc.} = 0,$$

*dico in ofcillationibus vniformibus fore vbique diftantiam*
*puncti M a linea verticali aequalem,*

$$\left(1 - \frac{x}{n} + \frac{xx}{4nn} - \frac{x^3}{4.9\,n^3} + \frac{x^4}{4.9.16\,n^4} - \text{etc.}\right)\beta : \left(1 - \frac{l}{n} + \frac{ll}{4nn}\right.$$
$$\left. \frac{l^3}{4.9.n^3} + \frac{l^4}{4.9.16\,n^4} - \text{etc.}\right)$$

*adeoque diftantiam puncti infimi* F *futuram effe aequalem*

$$\beta : \left(1 - \frac{l}{n} + \frac{ll}{4nn} - \frac{l^3}{4.9\,n^3} + \frac{l^4}{4.9.16\,n^4} - \text{etc.}\right).$$

*Erit porro longitudo penduli fimplicis ifochroni, vt antea*
$= n$, *feu in cafu fimpliciffimo proxime,* $=$

$$\frac{211 l^4 - 844 l^3\lambda + 1536 ll\lambda\lambda + 1440 l\lambda^3 + 576\lambda^4}{3041 l^4 + 9120 l^3\lambda + 1152 ll\lambda\lambda + 576\lambda^3}$$

Corol.

# Corollarium.

**23.** Sit v. gr. longitudo fili eadem quae longitudo catenae, id eft, $l = \lambda$, erit longitudo penduli fimplicis ifochroni feu $n$ proxime $= 1, 56\,l$: diftantiaeque punctorum extremorum F et N a linea verticali fe fere habebunt vt 11 ad 1: plures tamen praeter hunc alii cafus fatisfacient fimiles illis, quos in paragrapho 21. enumerauimus: ita poft dictum cafum fequitur is, quo fit $n$ fere $= \frac{1}{5}l$; punctumque C excurfiones contrarias facit cum puncto A atque triplo maiores.

# Theorema II.

Fig. 10.

**24.** *Pofitis omnibus vt in theoremate decimo, fi catena in origine A pondere onerata fuerit tanto, quantum ineft catenae parti longitudinis* L; *erunt omnia vt in eodem theoremate decimo, fi modo nunc fiat*

$$1 - \frac{(L\lambda + Ll + ll + l\lambda)}{n(L+l)} + \frac{(4\,ll\lambda + l^3 + 2\,ll\lambda)}{4\,nn.(L+l)} - \frac{(9\,lll\lambda + Ll^3 + l^4 + 3\,l^3\lambda)}{4.\,9n^3(L+l)}$$
$$+ \frac{(16\,l^3 + L\lambda + Ll^4 + l^5 + 4\,l^4\lambda)}{4.9.16\,n^4(L+l)} - etc. = 0.$$

*Fuerit v. gr.* $L = \lambda = l$, *et fiet* $1 - \frac{2l}{n} + \frac{11}{nn} - \frac{7\,l^3}{36\,n^3} + \frac{11\,l^4}{576\,n^4} - etc. = 0$, *hincque habebitur proxime* $n = 1, 37\,l$; *arculi autem a punctis catenae infimo et fupremo defcripti erunt vt* 100 *ad* 39.

# Theorema 12.

## GENERALE PRO CATENIS VTCVNQVE INAEQVALITER CRASSIS ATQVE GRAVIBVS.

**25.** Fuerit denique catena vtcunque inaequalis ftructurae ita vt pofito longitudine partis catenae $FM = x$,

*fit*

## DE OSCILLATIONIBVS CORPORVM. 121

*fit pondus eius $\xi$, intelligendo per $\xi$ qualemcunque fun-
&ionem ipfius x: vocetur porro diftantia puncti M ad libitum affumti a linea verticali $=y$: dico curuaturam F
MN hac definiri aequatione, fumta dx pro conftante, $\int y$
$d\xi = -\frac{n\xi dy}{dx}$: huicque aequationi poftquam in quolibet cafu particulari recte fatisfactum fuerit, fore longitudinem
penduli fimplicis ifochroni $= n$.*

## Corollaria.

26. I. In catenis aequabilis craffitiei quarum pondus
integrum $= 1$, eft $\xi = \frac{x}{l}$: pro his igitur talis inferuit
aequatio $\int y\, dx = -\frac{nx\, dy}{dx}$, ex qua omnia deduci poffunt,
quae a paragrapho decimo fexto ad vigefimum tertium
dicta funt.

II. Fuerit pondus catenae integrae rurfus $= 1$:
longitudo eius $= l$: fitque vbique $\xi = \frac{xx}{ll}$; erunt diftantiae punctorum F et M a linea verticali vt 1 ad fum-
mam huius feriei

$$1 - \frac{x}{n} + \frac{xx}{3nn} - \frac{x^3}{3.6\,n^3} + \frac{x^4}{3.6.10\,n^4} - \frac{x^5}{3.6.10.15\,n^5} + \text{etc.}$$

Eft autem $n$ longitudo talis, vt fit

$$1 - \frac{l}{n} + \frac{ll}{3nn} - \frac{l^3}{3.6\,n^3} + \frac{l^4}{3.6.10\,n^4} - \frac{l^5}{3.6.10.15\,n^5} + \text{etc.} = 0.$$

cui conditioni proxime fatisfit, cum fumitur $n = \frac{11}{20} l$;
tantaque eft longitudo penduli fimplicis ifochroni: erit
autem excurfio puncti infimi F fere tripla eius quam
facit punctum medium M.

## Scholium Generale.

27. In omnibus quas confiderauimus ofcillationibus diftantiae fingulorum punctorum a linea verticali,

*Tom. VI.*                    Q                    quafi

## 122 *THEOREMATA DE OSCILLAT. CORPOR.*

quafi infinite paruae cenfendae funt ratione longitudinis fili corpora connectentis aut catenae, imo etiam
ratione arcuum catenae, de quibus in paragrapho 21.
diximus, ita vt v. gr. in figura feptima etiam diftantia F C infinities minor effe debeat linea C B, ad quod
animus eft aduertendus in inftituendis experimentis,
quamuis a magnitudine ofcillationum non facile error
admodum notabilis oriatur.

Si haec pendula in turbinem agantur, eandem
figuram induent, quam ipfis in ofcillationibus affigna-
uimus, et gyros fuos duplo abfoluent tempore, quo
ofcillationes in eodem perficiunt plano.

Comment: Acad: Sc: Tom: VI. Tab: VII. p. 10.

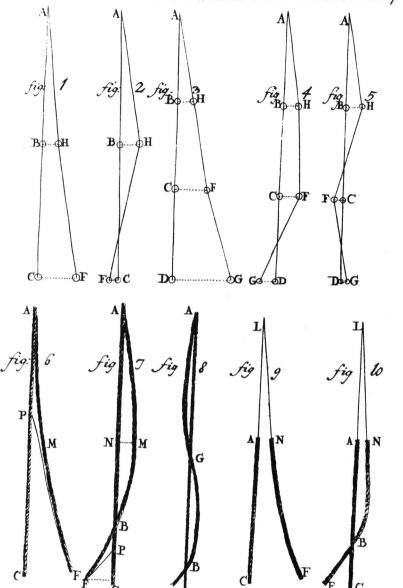

# COMMENTARII
## ACADEMIAE
## SCIENTIARVM
# IMPERIALIS
## PETROPOLITANAE.

TOMVS VII.

AD ANNOS cɔlɔccxxxiv. & cɔlɔccxxxv.

PETROPOLI,

TYPIS ACADEMIAE. cɔlɔccxl.

162    *DE OSCILLATIONIBVS CORPORVM*

## *Danielis Bernoulli*
# DEMONSTRATIONES
# THEOREMATVM SVORVM
## DE
# OSCILLATIONIBVS CORPORVM
### FILO FLEXILI CONNEXORVM ET CATENAE
### VERTICALITER SVSPENSAE.

## I.

**Tabula IX** **D**Edi nuper theoremata de oscillationibus corporum filo flexili connexorum: demonstrationes autem, quas tum non vacabat in ordinem redigere, nunc paullo plus otii nactus eo libentius cum publico communicabo, quod multorum aliorum similium problematum solutio inde peti possit, eorum praesertim in quibus motus partium non sunt inter se paralleli. Inter huiusmodi problemata facillimum est atque a multis iam diu solutum, quod circa centra oscillationis inuenienda versatur. Ad ea quoque pertinet problema de motu mixto determinando, quo corpus ex pluribus diuersae grauitatis specificae partibus compositum in fluido descendit: pertinent porro theoremata, quae in Commentar. Tom. II. p. 200. a Patre cum publico communicata fuerunt: Praesertim autem methodus, quam mox exhibebo, cum successu adhibetur, quando in systemate corporum plurium lege aliqua inter se connexorum, situs vnius ex situ alterius cognito non potest immediate determinari,

## FILO FLEXILI CONNEXORVM &c.    163

nari, veluti cum corpus super hypothenusa trianguli in
horizonte mobilis descendit; hic enim si vel noueris
situm corporis in hypothenusa, ipsius tamen trianguli
situs in horizonte incognitus manebit nisi hunc aliunde
determinare scias. Problema hoc postremum aliquando
Patri meo proposueram et plane inter se conuenerunt
solutiones nostrae; Eam, quae a Patre profecta est,
Academia Commentariis suis inseri curauit, vid. Tom. V.
p. 11. Quae ad hanc classem pertinent, nouam me-
chanicae partem efficiunt: Principium autem, quo vti so-
leo ad huiusmodi problemata soluenda, tale est:

Puta in systemate ad momentum temporis corpora
singula a se inuicem solui, nulla facta attentione ad mo-
tum iam acquisitum, quia hic de acceleratione seu mutatione motus elementari tantum sermo est: ita quolibet
corpore situm suum mutante, systema aliam accepit figuram, quam non-solutum habere debebat: Igitur finge
causam mechanicam quamcunque systema in debitam
figuram restituentem atque, rursus inquiro in mutationem
situs ab hac restitutione ortam in quolibet corpore; et
ex vtraque mutatione intelliges mutationem situs in systemate non soluto, indeque accelerationem retardatio-
nemue veram cuiusuis corporis ad systema pertinentis
obtinebis.

Quomodo haec regula ad praesens nostrum de oscillationibus corporum filo flexili ligatorum aut catenae
verticaliter suspensae determinandis, negotium applican-
da sit, hic docebo, alia occasione idem fortasse etiam

X 2                                            mon-

## 164  *DE OSCILLATIONIBVS CORPORVM*

monſtraturus in problematis aliis partim iam a Patre meo tractatis partimque nouis.

Figura 1.     II. Sit filum A H F grauitatis expers, duobus oneratum ponderibus in H et F, e puncto fixo A ſuſpenſa: faciant corpora oſcillationes veluti infinite paruas, ſintque eorundem diſtantiae a linea verticali A C, vt $2\,\mathrm{M}\,l$ ad

$$m\,\mathrm{L} - m\,l + \mathrm{M}\,\mathrm{L} + \mathrm{M}\,l \pm V(\,4\,m\,\mathrm{M}\,\mathrm{L}\,\mathrm{L} + (m\,l + \mathrm{M}\,\mathrm{L} + \mathrm{M}\,l - m\,\mathrm{L}\,)^2\,).$$

Demonſtrandum eſt, oſcillationes in vtroque corpore fore iſochronas. Valores litterarum $m$, M, $l$ et L infra dabuntur.

Erunt oſcillationes iſochronae, ſi fuerint vires acceleratrices in corporibus vt diſtantiae eorundem a linea verticali; nec enim differunt diſtantiae hae ab arcubus deſcribendis: Has igitur vires acceleratrices definiemus: ponatur pars fili H F extendi facillime, ita vt corpus F nihil amplius retineat: accelerabitur corpus iſtud verticaliter deorſum a grauitatis vi naturali: finge ita accelerari vt perueniat dato tempuſculo ex F in E, dum eodem temporis puncto alterum corpus filo A H alligatum arculum H L deſcribit: ductae iam intelligantur horizontales L B et E C, quae quamuis ceu infinite paruae conſiderentur, ſint tamen arculo L H infinities maiores. Apparet ex mechanicis et theoria infinite paruorum, fore $\mathrm{H\,L} = \frac{\mathrm{B\,L}}{\mathrm{L\,A}} \times \mathrm{F\,E}$. Poſitis igitur corporibus in L et E ductiſque rectis A L et L E, erit quidem filum A L inuariatae longitudinis, alterum autem L E iam maioris erit longitudinis quam fuerat in ſitu H F: concipiatur igitur cauſa, quae filum L E contrahat ad ſuam

lon-

## FILO FLEXILI CONNEXORVM &c.  165

longitudinem naturalem: dico ab ifta contractione corpus ex E eleuatum iri vsque in $u$, alterumque retractum
ex L in $n$: fpatiola E$u$, L$n$ determinabimus, poftquam
monuero, quod, ducta minima recta F$u$, verae accelerationes, quae durante affumto tempusculo acceffereunt,
feu ipfae etiam vires acceleratrices rationem habiturae
fint in corporibus H et F vt H$n$ et F$u$. Sed vt ratio intelligatur inter H$n$ et F$u$, faciemus AH feu AL
$= l$: HF feu $nu = $ L: maffa in corpore fuperiori $= m$;
in inferiori $= $ M. Producatur AL et ex E in illam
perpendicularis ED demittatur. Denique ducantur horizontalis HG et verticalis FG: erit F$u$ ad $nu$ perpendicularis cenfenda atque triangulum minimum FE$u$ trian-
gulo HFG fimile, ipfaque E$u$ lineolae FE aequalis:
vnde fi ponatur BL $= $ 1; DE $= x$; erit MC $= 1 + \frac{L}{l}$;
EC $= x + 1 + \frac{L}{l}$; HG $= $ CE $-$ BL $= x + \frac{L}{l}$; hinc

$$F u = (\tfrac{1}{l} + \tfrac{x}{L}) \times F E:$$

Supereft vt definiatur H$n$: Notetur quod filum LE,
dum contrahitur, corpus E directe furfum trahit; dum
corpus alterum L oblique ad directionem fuam L$n$ re-
trahit: hoc igitur titulo erit L$n$ ad E$u$ vt DE ad LE
feu vt $x$ ad L: fed eft praeterea L$n$ ad E$u$ reciproce
vt maffa corporis L ad maffam corporis E, id eft, directe vt M ad $m$: compofita ratione erit L$n$ : E$u =$ M$x$ :
$m$L; vnde pofita FE pro E$u$, erit L$n = \frac{M x}{m L} \times$ FE; et
quia H$n = $ HL $-$ L$n$, fequitur fore

$$H n = (\tfrac{1}{l} - \tfrac{M x}{m L}) \times F E:$$

funt igitur vires acceleratrices in corporibus H et F,

vt

## 166   DE OSCILLATIONIBVS CORPORVM

vt $\frac{1}{l}+\frac{x}{L}$ ad $\frac{1}{l}-\frac{Mx}{mL}$: ponantur hae vires ad ifochronismum obtinendum proportionales fpatiis defcribendis L B et EC, feu fiat $(\frac{1}{l}+\frac{x}{L}):(\frac{1}{l}-\frac{Mx}{mL})=1:(x+1+\frac{L}{l})$, atque reperietur facta reductione

$$x=\frac{mL-ml-ML-Ml+\sqrt{[4mMLL+(ml+ML+Ml-mL)^2]}}{2ML}$$

Huic autem fi addatur M C feu $1+\frac{L}{l}$, habebitur

$$CE=\frac{mL-ml+ML+Ml\mp\sqrt{[4mMLL+(ml+ML+Ml-mL)^2]}}{2Ml}\times BL,$$

plane vt habet in parte huius argumenti prima theorema tertium Prop. 7.

III. Pofitis iisdem, erit longitudo penduli fimplicis ifochroni aequalis

$$\frac{2mLl}{mL+ml+Ml+ML\mp\sqrt{[4mMLL+(ml+ML+Ml-mL)^2]}},$$

cuius rei rationem intelliges ex eo, quod fi pendulum fimplex longitudinis A H feu $l$ confideretur, fit vis acceleratrix in hoc pendulo fimplici ad vim acceleratricem corporis H in pendulo noftro compofito vt H$l$ ad H$n$, id eft, vt $\frac{1}{l}$ ad $\frac{1}{l}-\frac{Mx}{mL}$: funt autem longitudines pendulorum ifochronorum in reciproca ratione virium acceleratricium; Erit igitur longitudo penduli quaefiti ad longitudinem A H vt $\frac{1}{l}$ ad $\frac{1}{l}-\frac{Mx}{mL}$: vnde inuenitur longitudo penduli ifochroni $=\frac{mLl}{mL-Mlx}$, et pofito valore pro $x$ fupra inuento, erit eadem longitudo, vt dictum eft, aequalis

$$\frac{mLl}{mL+ml+M+NL\mp\sqrt{[4mMLL+(ml+ML+Ml-mL)^2]}}$$

**Figura 2.**   IV. Si filum A G fit tribus oneratum corporibus in H, F et G, ofcillationes facientibus valde paruas et ifochro-

## *FILO FLEXILI CONNEXORVM &c.*   167

chronas, ponaturque maffa corporis fupremi $= m$; medii $= M$ et infimi $= \mu$: diftantia $AH = l$; $HF = L$;
$FG = \lambda$: diftantia corporis H a linea verticali $AP = I$;
diftantia vero corporis F ab eadem linea verticali $= s$;
erit

$$((MMl\lambda + M\mu l\lambda)ss + mMl\lambda + m\mu lL - mML\lambda$$
$$- MMl\lambda - MMLλ + m\mu\ \lambda - M\mu l\lambda - M\mu L\lambda)s$$
$$- m\mu l\lambda - mMl\lambda) \times ((Ml\lambda + \mu l\lambda)s - mL\lambda - Ml\lambda$$
$$- MLλ - \mu l\lambda - \mu L\lambda + mlL) = mm\mu llLLs.$$

Diftantia autem corporis infimi a linea verticali erit
pro quauis radice ipfius $s$ aequalis

$$(\tfrac{MM\lambda}{m\mu L} + \tfrac{M\lambda}{mL})ss + (1 + \tfrac{\lambda}{L} + \tfrac{M\lambda}{\mu L} - \tfrac{M\lambda}{\mu l} - \tfrac{MM\lambda}{m\mu L} - \tfrac{MM\lambda}{m\mu l} - \tfrac{M\lambda}{mL} -$$
$$\tfrac{M\lambda}{ml})s - \tfrac{M\lambda}{\mu L} - \tfrac{\lambda}{L}.$$

Haec vt demonftrentur, ponatur rurfus filum infimum FG facillime extendi atque fic corpus G vi grauitatis naturali acceleratum, aflumto aliquo tempusculo
infinite paruo verticaliter defcendere ex G in $s$, dum
interea ambo corpora fuperiora accelerentur, vti in figura prima, faciendo arculos fuos $Hn$ et $Fu$. Patet autem, fi $Gs$ in figura fecunda aequalis ponatur defcen-
fui FE in figura prima, fore pariter arculos $Hn$ et $Fu$
idem in vtraque figura; erit igitur per praecedentem pa-
ragraphum $Hn = (\tfrac{1}{l} - \tfrac{Mx}{mL}) \times Gs$ et $Fu = (\tfrac{1}{l} + \tfrac{x}{L} \times Gs$, intelligendo per $x$ lineolam $uM$ perpendiculariter ad pro-
longatam $An$ ductam, prouti deinceps per $y$ intellige-
mus lineolam $yv$, quae perpendicularis eft ad prolon-
gatam $nu$: iam ducantur horizontalis FQ ac verticalis QG,
fumtaque

## 168   *DE OSCILLATIONIBVS CORPORVM*

fumtaque $uy = FG$, ducatur $Gy$. His ita ad calculum praeparatis, nunc rurfus fingendum eft, rectam $us$, in priftinam longitudinem $FG$ contrahi: ita eleuabitur corpus ex $s$ in $y$ vel in $r$ (eft autem $yr$ nulla prae $Gy$); corpora autem fuperiora iterum retrahentur ex $n$ in $o$ et ex $u$ in $m$: atque fic tandem patet fore vires acceleratrices in fingulis corporibus fecundum directiones fuas naturales ad vim grauitatis naturalem, vt fe habent $Ho$, $Fm$ et $Gy$ ad $Gs$: fupereft igitur vt fingula haec elementa exprimantur, probe obferuato arculos $Hn$, $Fu$ etc. nullos effe prae diftantiis corporum a linea verticali. Inuenietur autem recte inftituto calculo $FL = \frac{\lambda}{l}$ $+\frac{\lambda}{L}x + y$: et quia $FG:FQ = Gs:Gy$, erit

$$Gy = \left( \frac{1}{l} + \frac{x}{L} + \frac{y}{\lambda} \right) \times Gs.$$

Iam porro quaerendum eft, quantae futurae fint retrogradationes corporum in $u$ et $n$ pofitorum, quae fiunt, dum corpus infimum ex $s$ in $y$ aut in $r$ eleuatur. Notetur potentiam filum $us$ contrahentem vbique aequaliter diffundi. Erit igitur rurfus vt in fuperiori paragrapho $um$ ad $sy$ feu ad $Gs$ in ratione compofita ex $vy$ ad $uy$ et maffae $\mu$ ad maffam $M$: vnde inuenitur $um$ $= \frac{\mu y}{M\lambda} \times Gs$, qua fubtracta ab $Fu$ feu ab $\left( \frac{1}{l} + \frac{x}{L} \right) \times Gs$, oritur

$$Fm = \left( \frac{1}{l} + \frac{x}{L} - \frac{\mu y}{M\lambda} \right) \times Gs.$$

Denique quia ab eo, quod corpus medium ex $u$ in $m$ cedit, nihil patitur corpus fupremum, erit, vt antea, $no$ ad $ys$ feu $Gs$ in ratione compofita ex $Mu$ ad $un$ et maffae $\mu$ ad maffam $m$; vnde $no = \frac{\mu x}{ML} \times Gs$: hacque fub-

## FILO FLEXILI CONNEXORVM &c. 169

fublata ab $n$H feu ab $(\frac{1}{l}-\frac{Mx}{mL}) \times$ G $s$, fiet.

$$H o = (\frac{1}{l} - \frac{Mx}{mL} - \frac{\mu x}{mL}) \times G s.$$

Poſtquam ſic accelerationes corporum ſingulorum in ve-
ris ſuis directionibus inuenimus, erunt hae diſtantiis ſuis
ab linea verticali $y$P, $u$C et $n$B ſeu quantitatibus ( $1 +$
$\frac{L}{l} + \frac{\lambda}{l} + x + \frac{\lambda}{L}x + y$), ($1 + \frac{L}{l} + x$) et ($1$) proportionales faciendae iſochronismi ergo: Ita duae ae-
quationes obtinebuntur valores $x$ et $y$ determinantes:
atque ſi deinde ponatur $1 + \frac{L}{l} + x = s$, inuenietur
aequatio pro $s$ eadem, quam ſupra recenſuimus, quam-
que demonſtrandam ſuſcepimus.

V. Acceleratio corporis H expreſſa per H$o$ ſeu per
$(\frac{1}{l} - \frac{Mx}{mL} - \frac{\mu x}{mL}) \times$ G $s$ eſt ad accelerationem eiusdem corporis, abſentibus duobus inferioribus expreſſam per $\frac{1}{l} \times$ G $s$,
vt $\frac{1}{l} - \frac{Mx}{mL} - \frac{\mu x}{mL}$ ad $\frac{1}{l}$, ſeu vt $mL - M lx - \mu lx$ ad $m$L.
Sequitur inde longitudinem penduli iſochroni eſſe.

$$\frac{mLl}{mL - Mlx - \mu lx}$$

Hanc autem non differre ab illa, quae in parte prima, propoſitione decima tertia data fuit, videbis ſi ibi,
prouti factae a nobis denominationes poſtulant, intelli-
gas per $x$, quod hic per $s$ ſeu per $1 + \frac{L}{l} + x$.

VI. Sint iam plura et quotcunque volueris corpora, Figura 3.
veluti B, C, D, E, F producantur ſingula fila deſignen-
turque ſinus angulorum BAN(AN eſt verticalis) CBG,
DCH, EDL, FEM per $p$, $q$, $r$, $s$, $t$; maſſae autem
corporum per ipſas litteras iisdem appoſitas denotentur,
*Tom. VII.*         Y         dico

170 *DE OSCILLATIONIBVS CORPORVM*

dico pofita **vi** grauitatis naturali $= 1$, vires acceleratrices corporum fecundum fuas directiones fore vt fequitur.

$$\text{in } B = p - \frac{C+D+E+F}{B} q,$$
$$\text{in } C = p + q - \frac{D+E+F}{C} r,$$
$$\text{in } D = p + q + r - \frac{E+F}{D} s,$$
$$\text{in } E = p + q + r + s - \frac{F}{E} t,$$
$$\text{in } F = p + q + r + s + t;$$

Veram hanc effe virium acceleratricium legem, percipies fi fextum corpus fuo filo inferius adhuc appendi ponas, calculumque deinde inftituas, vt fecimus ratione trium corporum paragrapho quarto, fingendo fcilicet, corpus infimum naturali grauitatis vi verticaliter deorfum accelerari, reliquis interim fecundum fuam indolem vibratis, idemque corpus mox a contractione fili iterum eleuari: ita enim legem hanc accelerationum nunc expofitam a quinque corporibus ad fex, et inde ad feptem atque fic quousque libuerit recte continuari videbis.

Ex affumtis autem fingulorum angulorum finibus, deducuntur corporum a linea verticali diftantiae, ac fi quamuis diftantiam vi acceleratrici refpondenti proportionalem facias, habebis tot aequationes quot incognitas, fic vt omnia denique defiderata inde recte definiri poffint.

VII. Puta nunc corpora effe numero infinita et
Figura 4 aequalia, diftantiis minimis et aequalibus a fe invicem pofita: ideam habebis catenae vniformis ab vna extremitate fuspenfae, qualis eft AC vel AF: In hac elemen-

## FILO FLEXILI CONNEXORVM &c.    171

mentum confideretur infinite paruum $Mm$ vel $Nn$, ductis $MN$ et $mn$ ad $AC$ perpendicularibus et $mo$ eidem $AC$ parallela: ponatur $Am$ vel $An = s$ (nec enim differunt quia infinite parum diftant); $mM$ vel $nN = ds$, quod elementum conftans affumatur: longitudo catenae integrae $AF = l$; $mn = y$; $Mo = dy$: erit (pofita vi grauitatis naturali $= 1$) per praecedens theorema vis acceleratrix in $m$ aequalis fummae omnium finuum angulorum contactus, qui funt inter $A$ et $m$, diminutae tertia proportionali corpusculi in $m$, fummae omnium corpusculorum in $MF$ et finus anguli contactus in $M$: fic igitur habetur vis acceleratrix in $M = \int \frac{d\,dy}{ds} - \frac{(l-s)\,d\,dy}{ds^2}$. Quia vero ifochronismus poftulat, vt vis acceleratrix fit proportionalis applicatae $MN$, erit affumta $n$ pro conftante $\int \frac{d\,dy}{ds} - \frac{(l-s)\,d\,dy}{ds^2} = \frac{y}{n}$: fumatur integrale termini primi fine additione conftantis, quia hic nulla fumenda eft: fic fiet $\frac{dy}{ds} - \frac{(l-s)ddy}{ds^2} = -\frac{y}{n}$. Denique ponatur $l-s$ feu $FM$ aut $CN = x$, et erit $\frac{-dy}{dx} - \frac{x\,ddy}{dx^2} = \frac{y}{n}$, fiue

$$n\,dy\,dx + n\,x\,ddy = -y\,dx^2,$$

quae aequatio denotat naturam curuae $AF$: quoniam vero integralis eius non apparet, pofui

$$y = \alpha - \beta x - \gamma xx - \delta x^3 - \varepsilon x^4 - \text{etc.},$$

$$dy = -\beta\,dx - 2\,\gamma\,x\,dx - 3\,\delta\,xx\,dx - 4\,\varepsilon\,x^3\,dx - \text{etc.}$$

$$ddy = -2\,\gamma\,dx^2 - 2.3.\delta\,x\,dx^2 - 3.4.\,\varepsilon\,xx\,dx^2 - \text{etc.}$$

Hisque valoribus fubftitutis diuifaque deinde aequatione per $dx^2$, oritur

Y 2

$-\beta$

$$-\mathcal{B}-2\gamma x-3\delta xx-4\varepsilon x^3-\text{etc.}$$
$$-2\gamma x-2.3\delta xx-3.4\varepsilon x^3-\text{etc.}=0,$$
$$+\frac{\alpha}{n}-\frac{\mathcal{B}}{n}x-\frac{\gamma}{n}xx-\frac{\delta}{n}x^3-\text{etc.}$$

cui aequationi fatisfit ponendo $\alpha=1$; $\mathcal{B}=\frac{1}{n}$; $\gamma=\frac{-1}{4nn}$;
$\delta=\frac{1}{4.9n^3}$; $\varepsilon=\frac{-1}{4.9.16n^4}$ etc. vnde

$$y=1-\frac{x}{n}+\frac{xx}{4nn}-\frac{x^3}{4.9n^3}+\frac{x^4}{4.9.16n^4}-\text{etc.}$$

vbi per 1 intelligenda eft diftantia puncti infimi F a
verticali: et quia pofita $x=l$, eft $y=0$ erit fimul

$$1-\frac{l}{n}+\frac{ll}{4nn}-\frac{l^3}{4.9n^3}+\frac{l^4}{4.9.16n^4}-\text{etc.}=0;$$

Hinc deriuandus eft valor litterae $n$, qui exprimet longitudinem fubtangentis in F. Haec demonftrant veri-
tatem theorematis, quod in praecedente differtatione
octauum eft.

VIII. Vt habeatur longitudo penduli ifochroni,
quaerenda eft vis acceleratrix in puncto F, quae per
§. VI. erit aequalis fummae finuum omnium angulorum
contactus ab A vsque in F, id eft $=\int\frac{-ddy}{dx}$ feu $=\frac{-dy}{dx}$;
ponendo fimul $x=0$; et hinc fit $\frac{-dy}{dx}=\frac{-1}{n}$. Eft itaque
vis acceleratrix in F ad vim acceleratricem naturalem
vt 1 ad $n$: fi vero pendulum fimplex longitudinis $l$
habeatur, erit vis acceleratrix in illo $=\frac{1}{l}$ fub eadem
diftantia à linea verticali; ergo vis acceleratrix in extremitate catenae eft ad vim acceleratricem in pendulo fimplici eiusdem longitudinis, vt $l$ ad $n$: Hincque
erit longitudo penduli fimplicis cum catena fimul vi-
brantis $=n$, vt habet theorema nonum in praemiffa
differtatione.

IX.

## FILO FLEXILI CONNEXORVM &c.   173

IX. Theoremata autem decimum et vndecimum vnice pendent à debitae conſtantis additione, eaque proinde ceu nimis nunc facilia hic non attingam: ſed duodecimum ex §. VI rurſus, hunc in modum deducetur.

Corpuscula nunc conſiderentur infinita et aequali‑ bus diſtantiolis à ſe inuicem poſita, ſed inaequalis ponderis: ita habebitur idea catenae pro lubitu inaequaliter craſſae; ſit haec ita formata, vt longitudini F M ($x$) pondus reſpondeat $\xi$, denotante $\xi$ fundtionem qualemcunque ipſius $x$: Erit (per §. VI.) vis acceleratrix in $M = \int \frac{-ddy}{dx} - \frac{\xi ddy}{a\xi dx} = \frac{y}{n}$, vel quia $dx$ conſtans, erit $\frac{-dy}{dx} - \frac{\xi ddy}{d\xi dx} = \frac{y}{n}$, aut $nd\xi dy + n\xi ddy = -yd\xi dx$, vel demique

$$\frac{-n\xi dy}{dx} = \int y\,dx,$$

vt fert theorema duodecimum, de quo ſermo erat: demonſtratio magis fiet intelligibilis, ſi ſimul conferatur paragraphus ſeptimus.

Y 3                                    DE

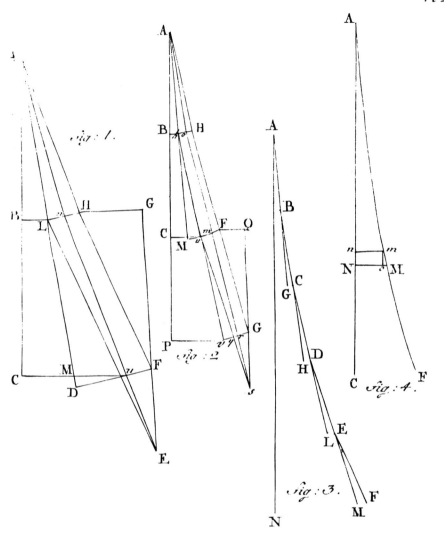

Commerc. Acad. Sc. Tom VII. Tab IX p.
162.

*Danielis Bernoulli*
# THEOREMATA DE OSCILLA-
## TIONIBVS CORPORVM FILO FLEXILI CON-
## NEXORVM ET CATENAE VERTICA-
## LITER SVSPENSAE

# Introduction to the Argument

T heories of oscillation for extended objects which authors have given up to the present time presuppose that the relative positions of the parts are fixed so that they are all carried with a common angular motion. But bodies suspended from a flexible thread require another theory. The principles usually used in mechanics do not seem to be sufficient for this business, since, of course, the relative positions of the bodies are variable and continually changing. A chain hanging vertically and agitated with an oscillatory motion once gave me an opportunity for thinking about these things. Seeing that the chain is thrown with an exceedingly irregular motion it occurred to me to consider how the chain should be curved so that if all its parts begin their motion simultaneously they will arrive simultaneously at the vertical line going through the point of suspension. I realized that oscillations of this kind would be regular and such that the times [of oscillation] could be determined. Soon, however, I found that this curve is difficult to determine unless one begins the investigation with the simplest cases. And so I began these considerations with two bodies held together a given distance apart by a flexible thread; then I considered three, and soon four, and finally any number of bodies any distance apart. And when I made the number of bodies infinite, I saw at last the nature of the oscillating chain whether equal or unequal in thickness but everywhere perfectly flexible. Let me go through each successive case. But I shall reserve for another occasion demonstrations which are not worth furnishing now. To find the solution I used new principles, and therefore I wanted to confirm the theorems with experiments, so that even people who have not been able to give their full attention to these esoteric matters (and who could thus easily come upon a false solution) would have no doubt about their truth.

We will not consider oscillations other than those that are minimal and isochronous; for the experiments, however, it will be possible to use somewhat larger ones without noticeable error.

# Theorem 1

2. *A perfectly flexible weightless string* AHF *is suspended from the point* Fig. 1, 2
A. *Let two equal weights be attached to it at H and at F: the lower weight is as far from the higher weight as the higher weight is from the point of suspension. Next let* ABC *be the vertical, and let the displacements of the weights H and F from this line be as though infinitely small. Finally, let the minimal horizontal lines* HB *and* FC *be drawn. I say that if both bodies begin to oscillate simultaneously, they will reach the vertical line at the same time, and in this way carry out their oscillations uniformly, provided one has chosen*

$$\mathsf{CF:BH} = 1 \pm \sqrt{2} : 1$$

# Corollary

3. Therefore uniform oscillations are made in two ways; when $\mathsf{CF} = (1 + \sqrt{2})\mathsf{BH}$, as the first figure shows, and also when $\mathsf{CF} = (1 - \sqrt{2})\mathsf{BH}$, as in the second figure.

# Theorem 2

4. *When the oscillations of the bodies* H *and* F *are made uniformly, the length of the simple tautochronous pendulum will be* $= (1/(2 \pm \sqrt{2}))$AH *or* $(1/(4 \pm \sqrt{8}))$AC, *where the plus sign applies to the contrary oscillations of the second figure, the negative sign for the conspiring oscillations of the first figure.*

# Corollary

5. And so contrary oscillations are completed much more quickly than conspiring ones: for you will count [241] contrary oscillations while the conspiring ones complete a hundred. But the conspiring oscillations differ less [than the contrary ones] from those that would be made under the same circumstances if the thread AHF were rigid. Nevertheless, bodies on a rigid thread oscillate a little more quickly than those on a flexible one, for the number of oscillations completed in equal times will be a little bit more than 1012 to 1000.

# Scholium

6. So that I might subject these propositions to experiment, I used two perfectly similar lead balls which were perforated [because] while they were being poured I restricted them in the middle to make a hole. The balls were connected by means of knots to a steel thread passing through in such a way that the low one was twice as far from the point of suspension as was the higher one. With my fingers on the tense thread I drew the lower ball to the position F. Soon the oscillations became uniform and with the help of divisions marked on the wall I ascertained that the displacements of the bodies H and F in fig. 1 were as 100 to 241, that is, as 1 to $1 + \sqrt{2}$ (par. 2). Also the number of oscillations occurring in a given time corresponded accurately to the length of the simple isochronous pendulum $(1/(2 - \sqrt{2}))$AH defined in [par.] 4. Next, when $FC = (1 - \sqrt{2})$BH as in the second figure, I held the balls by hand at F and H and then released them at the same moment. Reasonably uniform oscillations arose as before when the other proportion of distances FC and HB was used. Again, the number of oscillations agreed accurately with that of the simple isochronous pendulum of length $(1/(2 + \sqrt{2}))$AH in [par.] 4.

# Theorem 3

## IN GENERAL FOR TWO BODIES

7. *Now suppose that the part of the thread* AH $= l$; HF $= L$; *the weight of the body* H $= m$, *and of the other* F $= M$. *I say that the oscillations will be uniform if*

$$\text{CF} = \frac{mL - ml + ML + Ml \pm \sqrt{(4mMLL + (ml + ML + Ml - mL)^2)}}{2Ml} \times \text{BH}.$$

*The length of the simple isochronous pendulum will be*

$$= \frac{2mLl}{mL + ml + Ml + ML \mp \sqrt{(4mMLL + (ml + ML + Ml - mL)^2)}}$$

*or, supposing* $L + l = \lambda$ *and* $M + m = \mu$,

$$= \frac{2m\lambda l - 2mll}{\mu \mp \sqrt{(\mu\mu\lambda\lambda - 4m\mu l\lambda + 4m\mu ll)}}.$$

# Theorem 4

Fig. 3,        8. *If there are three instead of two similar bodies separated from each*
4, 5      *other by the same distance as that which separates the highest one from the point of suspension* A, *uniform oscillations are possible in three different*

*ways. The first is that which the third figure indicates when, assuming* $\mathsf{BH} = 1$, *one takes* $\mathsf{CF} = 2.292$ *and* $\mathsf{DG} = 3.922$; *the second, which is represented by the fourth figure, is obtained by making* $\mathsf{CF} = 1.353$ *and* $\mathsf{DG} = -1.044$; *and the third when, as in the fifth figure,* $\mathsf{CF} = -0.645$ *and* $\mathsf{DG} = 0.122$. $\mathsf{CF}$ *is, of course, found by taking the three roots of the equation*

$$4x^3 - 12xx + 3x + 8 = 0,$$

*and then for any root one takes*

$$\mathsf{DG} = 2xx - 2x - 2.$$

# Theorem 5

9. *When uniform oscillations are made as described, the length of the simple isochronous pendulum will be, in the case of the third figure, approximately* $2.406\mathsf{AH}$; *in the case of the fourth figure,* $0.0436\mathsf{AH}$; *and in the case of the fifth figure,* $0.159\mathsf{AH}$. *The length of the isochronous pendulum is, obviously,* $(1/(5 - 2x)) \times \mathsf{AH}$, *assuming again* $4x^3 - 12xx + 3x + 8 = 0$.

# Scholium

10. I accurately confirmed both these theorems by experiment in the case of the third figure, drawing the lowest body away from the vertical line and then letting it go. Although in the first oscillations a certain inequality could be felt, this spontaneously and very quickly disappeared, so that the displacements of each body remained, as far as could be discerned by eye, the same in many successive vibrations. Moreover, when measurements were made of the same things, a proportion of such a kind as theorem 4 indicates was found between them. Also the number of oscillations corresponded perfectly to the fifth theorem. The two remaining cases require much work: nevertheless I was able to bring about oscillations of both types well enough that the truth of the fifth theorem was apparent.

# Theorem 6

## IN GENERAL FOR THREE BODIES

11. *Now, the weights of the bodies* $\mathsf{H}$, $\mathsf{F}$, $\mathsf{G}$ *are arbitrary, and also their distances from the point of suspension are in any proportion. Let the weight of the body* $\mathsf{H} = m$, *of the body* $\mathsf{F} = M$, *and of the body* $\mathsf{G} = \mu$; *and then*

$AH = l$, $HF = L$, and $FG = \lambda$; *I say that oscillations will be uniform if assuming* $BH = 1$, $[C]F = x$, *one has*

$$\{(MMl\lambda + M\mu l\lambda)xx + (mMl\lambda + m\mu lL - mML\lambda - MMl\lambda - MML\lambda$$

$$+ m\mu l\lambda - M\mu l\lambda - M\mu L\lambda)x - m\mu l\lambda - mMl\lambda\} \times \{(Ml\lambda + \mu l\lambda)x$$

$$- mL\lambda - Ml\lambda - ML\lambda - \mu l\lambda - \mu L\lambda + mlL\} = mm\mu llLLx$$

*and also one takes for any root*

$$DG = \left(\frac{MM\lambda}{m\mu L} + \frac{M\lambda}{mL}\right)xx + \left(1 + \frac{\lambda}{L} + \frac{M\lambda}{\mu L} - \frac{M\lambda}{\mu l} - \frac{MM\lambda}{m\mu L} - \frac{MM\lambda}{m\mu l} - \frac{M\lambda}{mL} - \frac{M\lambda}{ml}\right)x$$

$$- \frac{M\lambda}{\mu L} - \frac{\lambda}{L}.$$

# Corollaries

12. I. The mass $\mu$ of the lowest body is assumed equal to zero, and the fundamental equation of the above paragraph is divided by the second factor which has a non-useful root. The first factor is $= 0$, and hence one will have $Mlxx + (ml - mL - Ml - ML)x - ml = 0$, or

$$x = \frac{mL - ml + ML + Ml \pm \sqrt{(4mMll + (ml - mL - Ml - ML)^2)}}{2Ml};$$

Moreover, it is to be noted that this value does not differ from that which we gave in the third theorem although the quantities on the inside of the parentheses under the sign of the radical have different signs.

II. If the mass of the middle body indicated by $M$ is assumed $= 0$, then, in order to have agreement between the third and sixth theorems, $L + \lambda$ and $\mu$ in the latter theorem are to be understood as the quantities designated by $L$ and $M$ in the former; and the line $DG$ defined in the preceding paragraph is to be compared with the line $CF$ in the third theorem. Anyone who pays attention to this will discover by doing some calculations that the equations of both theorems are the same.

III. Finally, when the mass $m$ of the highest body $H$ is assumed equal to zero, either one of the factors can be assumed $= 0$, and in either way one obtains $CF$ or $x = 1 + (L/l)$, just as the nature of the thing requires, because then the lines $AH$ and $HF$, as is clear, must lie in one direction. But the displacement of the lowest body cannot be determined from the equation unless it is done by some special method. Nor can the oscillations be immediately determined by the sixth theorem, since one or another of the lengths $L$ or $\lambda$ vanishes when two bodies are united.

IV. In the fourth figure one can so arrange things so that $CF = 0$, in which case, since the distances of the bodies from the vertical line maintain the same proportion during the whole oscillation, the middle body F remains at rest while both the others are agitated back and forth. And consequently it is clear that the length of the isochronous pendulum will be $\lambda$, since the lowest body oscillates just as though suspended from a fixed point C; indeed, that case about which we speak is obtained by setting $x = 0$ or rather

$$[\lambda] = \frac{mlL}{mL + Ml + ML + \mu l + \mu L}.$$

# Theorem 7

## WHICH DETERMINES IN GENERAL THE TAVTOCHRONOVS PENDVLVM FOR THREE OSCILLATING BODIES

13. *Retaining the notation and equations of the sixth theorem, I say that the oscillations of each body will be isochronous with the oscillations of a simple pendulum whose length is* $mlL/[mL + (M + \mu)(l + L - lx)]$.

# Corollary

14. In the case $x = 0$, which we have selected [par. 12.IV.], the length of the isochronous pendulum is $\lambda = mlL/(mL + Ml + ML + \mu l + \mu L)$, which agrees with *Cor.* 4, *Theorem* 6. If in addition it is assumed that $L = l$, the length of the isochronous pendulum is $\lambda = ml/(m + 2M + 2\mu)$: This agrees with *Problem* 1 which my father gave in *Comm. Acad. Petrop.* Vol. III, p. 15. And it is obvious that the weight $P$ which in that problem is attached to one end of the string is here the sum of the weights $G$ and $F$ increased by half the weight $H$.

# General Scholium

15. I can give similar equations for four, five, and however many bodies one would like: but the equation always rises to as high a degree as the number of bodies, and it is generally quite lengthy. But because the final equation arises from a system of linear equations, the law appears from the method I have used, by means of which all things associated with the equation to be determined can be found by iteration.

# Theorem 8

## ON THE SHAPE OF THE VNIFORMLY OSCILLATING CHAIN

16. *Let a uniformly heavy and perfectly flexible chain* AC *be suspended*   Fig. 6 *from the point* A, *and let it be understood to make uniform oscillations. The*

*chain has the extreme position* AMF. *The length of the chain is* $l$; *the length of a part* FM $= x$, *and n is assumed to have one of its values, so that*

$$1 - \frac{l}{n} + \frac{ll}{4nn} - \frac{l^3}{4 \cdot 9 \cdot n^3} + \frac{l^4}{4 \cdot 9 \cdot 16 \cdot n^4} - \frac{l^5}{4 \cdot 9 \cdot 16 \cdot 25 \cdot n^5} + \text{etc.} = 0.$$

*It is assumed that the distance of the end point* F *from the vertical line is equal to* 1; *I say that the distance of an arbitrary point* M *from the same vertical line equals*

$$1 - \frac{x}{n} + \frac{xx}{4n^2} - \frac{x^3}{4 \cdot 9 \cdot n^3} + \frac{x^4}{4 \cdot 9 \cdot 16n^4} - \frac{x^5}{4 \cdot 9 \cdot 16 \cdot 25n^6} + \text{etc.}$$

## Scholium

17. By the method that I gave in *Comm. Acad. Petrop. Tom. V de resolutione aequationum sine fine progredientium*, one finds by a very short calculation that $n$ is approximately $0.691l$. Therefore, if, for example, the point M is in the middle of the chain, it will be distant from the vertical line by approximately two fifths or, more accurately, three hundred and ninety eight thousandths of the distance of the lowest point F from the same vertical line. However, $n$ has an infinite number of other values.

## Theorem 9

18. *Using the notation of the eighth theorem, I say that the length of the simple pendulum isochronous with the oscillating chain is equal to n or rather to the subtangent* CP *to the curve* AF *at the lowest point* F; *or approximately equal to six hundred and ninety one thousandths of the whole chain in the case of the sixth figure.*

## Corollary

19. In this mode the chain oscillates more slowly than the rigid rod of uniform thickness of the same length as the flexible chain. For the oscillations of [the rod] are isochronous with the oscillations of a simple pendulum having two thirds the length of the rod.

## Scholium 1

20. After I connected a large number of lead balls close together along a thread, as said in the sixth paragraph, I used it as a chain for doing an experiment. I suspended the thread loaded with the balls from a firm point;

and having drawn it aside by the end F and having let it go, I observed the ratio between the distances of the extreme point F and the midpoint M from the vertical line AC, the oscillations now being made uniformly, and I found the same ratio as that which is indicated in the seventeenth paragraph. I also observed that the number of oscillations agreed with the length of the simple isochronous pendulum which is given in the ninth theorem.

# Scholium 2

21. Since the equation given in the eighth theorem, that is,

$$1 - \frac{l}{n} + \frac{ll}{4n^2} - \frac{l^3}{4 \cdot 9 \cdot n^3} + \frac{l^4}{4 \cdot 9 \cdot 16 n^4} - \frac{l^5}{4 \cdot 9 \cdot 16 \cdot 25 n^5} + \text{etc.} = 0$$

has infinitely many real roots, the chain can be inflected in infinitely many ways to make uniform oscillations. However, $n$ takes on smaller and smaller values, so that in the end it almost vanishes. The length of the simple isochronous pendulum is always equal to $n$ or to the subtangent CP; whence the oscillations will in the end be made as though infinitely fast. All possible cases are given in the following: first, when the chain does not intersect the vertical line at any point other than the point of suspension, which is represented by the sixth figure and for which the length of the simple isochronous pendulum is $n = 0.691l$, as we saw above. Next, when the chain crosses the vertical line in an additional fixed point, as indicated in the seventh figure, where the already-mentioned point of intersection is B: in this case the length of the isochronous pendulum is $n = 0.13l$, and twenty three oscillations are completed in the time that ten are completed for the case of the sixth figure; for M the point of maximum displacement, CN corresponds to $0.47l$ and MN is approximately $\frac{2}{5}$FC. After this case comes that that is shown in the eighth figure, where the vertical line is crossed by the oscillating chain at two fixed points B and G. Next [comes the case] when three intersections are made, and so forth. The arcs subtended between adjacent points of intersection are longer corresponding to their being higher. Moreover, in a chain that can be taken as though infinitely long the [shape of the] highest arc does not differ perceptibly from the shape of a tense musical string since the weight of that arc is as nothing compared with the weight of the whole chain. Nor is it difficult from this theory to deduce a theory of musical strings which clearly agrees with those that Taylor and my father gave—the first in his work *de methodo incrementorum*, the other in *Comm. Acad. Sc. Petrop.* Vol. III. Experiment teaches that intersections similar to those in vibrating chains occur also in musical strings, in fact a small piece of paper placed at certain places on the string does not move when an adjacent string vibrates.

# Theorem 10

Fig. 9     22. *If the chain* AC *is suspended from a weightless thread* LA, *and the distance* [*to the bottom*] *of a point chosen arbitrarily is supposed to be* FM $= x$, *the distance of the highest point* N *from the vertical line is* $\beta$, *and n takes on one of its values so that*

$$1 - \frac{(l+\lambda)}{n} + \frac{(ll+2l\lambda)}{4n^2} - \frac{(l^3+3l^2\lambda)}{4\cdot 9n^3} + \frac{(l^4+4l^3\lambda)}{4\cdot 9\cdot 16n^4} - \text{etc.} = 0,$$

*I say that in uniform oscillations the distance of any point* M *from the vertical line equals*

$$\left(1 - \frac{x}{n} + \frac{xx}{4nn} - \frac{x^3}{4\cdot 9n^3} + \frac{x^4}{4\cdot 9\cdot 16n^4} - \text{etc.}\right)$$

$$\times \beta : \left(1 - \frac{l}{n} + \frac{ll}{4nn} - \frac{l^3}{4\cdot 9n^3} + \frac{l^4}{4\cdot 9\cdot 16n^4} - \text{etc.}\right)$$

*and thus that the distance of the lowest point* F *will be equal to*

$$\beta : \left(1 - \frac{l}{n} + \frac{ll}{4nn} - \frac{l^3}{4\cdot 9\cdot n^3} + \frac{l^4}{4\cdot 9\cdot 16n^4} - \text{etc.}\right).$$

*The length of the simple isochronous pendulum will be equal, as before, to n, or rather, in the simplest case, approximately*

$$\frac{211l^4 + 844l^3\lambda + 1536ll\lambda\lambda + 1440l\lambda^3 + 576\lambda^4}{304l^3 + 912ll\lambda + 1152l\lambda\lambda + 576\lambda^3}$$

# Corollary

23. If, for example, the length of the thread is the same as the length of the chain, that is, $l = \lambda$, the length of the simple isochronous pendulum or $n$ will be approximately $1.56l$; and the distances of the endpoints F and N from the vertical line will be almost as 11 to 1. In addition to this, however, there will be many other cases similar to those which we listed in paragraph 21: thus, after the case mentioned follows that in which $n$ is approximately $\frac{1}{5}l$; and the point C has displacements in the direction contrary to those of point A and three times greater.

# Theorem 11

Fig. 10     24. *With everything assumed to be as in the tenth theorem, if the chain is loaded at the top* A *with a weight equal to the weight of a part of length*

*L of the chain, everything will be again as in the tenth theorem if now*

$$1 - \frac{(L\lambda + Ll + ll + l\lambda)}{n(L+l)} + \frac{([Ll^2] + 4lL\lambda + l^3 + 2ll\lambda)}{4nn(L+l)} - \frac{(9llL\lambda + Ll^3 + l^4 + 3l^3\lambda)}{4 \cdot 9n^3(L+l)}$$

$$+ \frac{(16l^3[\cdot]L\lambda + Ll^4 + l^5 + 4l^4\lambda)}{4 \cdot 9 \cdot 16n^4(L+l)} - \text{etc.} = 0.$$

*If, for example, $L = \lambda = l$, and*

$$1 - \frac{2l}{n} + \frac{ll}{nn} - \frac{7l^3}{36n^3} + \frac{11l^4}{576n^4} - \text{etc.} = 0,$$

*one has $n = 1.371$ approximately; the small arcs described by the lowest and highest points of the chain will be as 100 to 39.*

# Theorem 12

## IN GENERAL FOR CHAINS OF ARBITRARY THICKNESS AND WEIGHT

25. *Finally consider a chain of non-uniform construction, let $\xi$ be the weight of a length of a part of the chain $FM = x$, understanding $\xi$ to be any function of that $x$; let the distance of the arbitrary point $M$ from the vertical line be $= y$. I say that the curve $FMN$ is defined from the equation $\int y \, d\xi = -n\xi \, dy/dx$, with $dx$ taken as a constant element; and after this equation is satisfied correctly in any particular case, the length of the simple isochronous pendulum will be $= n$.*

# Corollaries

26. I. In chains of uniformly changing thickness the entire weight of which $= 1$, one has $\xi = x/l$; for these the equation given by $\int y \, dx = -nx \, dy/dx$ suffices, from which all the things that have been said in paragraphs sixteen to twenty-three can be deduced.

II. The weight of the entire chain is again $= 1$; its length $= l$; and let $\xi = xx/ll$ everywhere. The distances of the points $F$ and $M$ from this vertical line will be as 1 to the sum of the series

$$1 - \frac{x}{n} + \frac{xx}{3nn} - \frac{x^3}{3 \cdot 6n^3} + \frac{x^4}{3 \cdot 6 \cdot 10n^4} - \frac{x^5}{3 \cdot 6 \cdot 10 \cdot 15n^5} + \text{etc.}$$

Moreover, $n$ is of such length that

$$1 - \frac{l}{n} + \frac{ll}{3nn} - \frac{l^3}{3 \cdot 6n^3} + \frac{l^4}{3 \cdot 6 \cdot 10n^4} - \frac{l^5}{3 \cdot 6 \cdot 10 \cdot 15n^5} + \text{etc.} = 0.$$

a condition that is approximately satisfied when $n = \frac{11}{20}l$; and this is the length of the simple isochronous pendulum. Moreover, the displacement of the lowest point F will be approximately three times that of the midpoint M.

# General Scholium

27. In all the oscillations that we have considered the distances of all the points from the vertical line are taken to be as though infinitely small in comparison with the length of the thread connecting the bodies or the length of the chain and, indeed, also in comparison with the arc of the chain about which we spoke in paragraph 21, so that, for example, in the seventh figure the distance FC must be infinitely small in comparison with the line CB; one should pay attention to this in doing experiments, although a very conspicuous error does not arise easily from too great an amplitude in the oscillations.

If these pendula are carried in a revolving motion, they take on the same shape as that which we have assigned to those in oscillations, and they complete their revolutions in twice the time that they would take to perform [semi] oscillations in a plane.

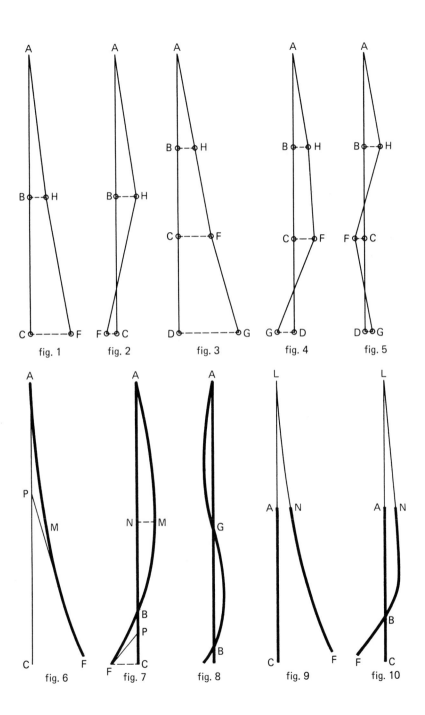

fig. 1  fig. 2  fig. 3  fig. 4  fig. 5

fig. 6  fig. 7  fig. 8  fig. 9  fig. 10

*Danielis Bernoulli*
DEMONSTRATIONES
THEOREMATVM SVORVM
DE
OSCILLATIONIBVS CORPORVM
FILO FLEXILI CONNEXORVM ET CATENAE
VERTICALITER SVSPENSAE

## I.

I recently gave some theorems on the oscillations of bodies that are connected by a flexible thread: there was not then time to give the demonstrations in detail; but having come upon a little more leisure, I will now more cheerfully communicate them to the public, because the solutions of many other similar problems, especially of those in which the motions of the parts are not parallel, can be sought by the same method. Among problems of this kind, the easiest one, already studied by many for a long time, concerns finding the center of oscillation. To this category belongs also the problem of determining the mixed motion with which a body made of many parts having different specific gravities descends in a fluid. Next belong the theorems which were communicated to the public by my father in *Comment. Tom.* II p. 200. However, the method which I will give presently can be used with success even when, in a system of many bodies connected to each other according to some law, the location of any single [body] cannot be immediately determined from the known location of another, as for example when a body descends along the hypotenuse of a triangle that is free to move horizontally. For in this case even if you know the place of the body on the hypotenuse, the location of the triangle on the horizontal will remain unknown unless you know how to determine this from another source. I once proposed this problem to my father, and our solutions agreed completely. (My father took care to put his solution in the *Acad. Comm.* See Vol. V, P. 11.) This class of problems takes one into a new domain of mechanics: the principle which I use to solve these problems is the following:

Suppose that in a system, at a moment of time, each body is in turn considered to be free, without paying attention to the motion already acquired, because one is here considering so elementary an acceleration or rather change in motion. But when any body changes its position, the system takes on a shape other than that which it would have had were the body not constrained. Therefore imagine some mechanical cause restoring the system to its original shape. I next inquire into the change of location of any body arising from this restitution. From the sum of the two changes you learn the change of place of a body in a constrained system, and you will thence obtain the true acceleration or retardation of any body that is part of the system.

I will show here how this rule is to be applied to our present problem of determining the oscillations of bodies connected by a flexible thread or of a vertically suspended chain. On another occasion, perhaps, the same thing will be shown in other problems, including some that were already treated by my father and some that are new.

Fig. 1 II. Let the weightless string AHF, loaded with two weights at H and F, be suspended from the fixed point A: Let the bodies make oscillations that can be considered to be infinitely small, and let their distances from the vertical line AC be as $2Ml$ to $mL - ml - ML + Ml \pm \sqrt{(4mMLL + (ml + ML + Ml - mL)^2)}$. It is to be shown that the oscillations of each body will be isochronous. The values of the letters $m$, $M$, $l$, and $L$ will be given below.

The oscillations will be isochronous if the accelerating forces on the bodies are as their distances from the vertical line since these distances do not differ from those described by the trajectories. Therefore we will determine these accelerating forces: Suppose first that the part of the string HF is easily extended, so that the body F will not be held back. That body will be accelerated downwards by the natural force of gravity. Imagine that it is accelerated so that it arrives at E from F in a given infinitesimal interval of time, while at the same point of time the other body attached to the string AH describes the arc HL. The horizontals LB and EC are understood to have been drawn; although considered as if infinitely small, they are nevertheless infinitely bigger than the arc LH. It is evident from mechanics and from the theory of the infinitely small that $HL = (BL/LA)FE$. Assume that the bodies are at L and E, and draw the lines AL and LE; the thread AL will, of course, be of invariant length, but the other, LE, will now be of greater length than it was at HF. Next, imagine that the cause which will contract LE to its natural length is activated. I say that by that contraction the one body will be raised from E to u, and the other drawn back from L to n: We will determine the elements of distance, Eu and Ln after I point out, having drawn the short line Fu, that the true accelerations that the bodies accumulate during the time interval, or rather the actual accelerating forces on the bodies H and F, will be proportional to Hn and Fu. But to find the ratio between Hn and Fu, we shall let AH or AL = $l$,

HF or nu $= L$, the mass of the upper body $= m$, of the lower $= M$. Let AL be extended and let the perpendicular ED be dropped to it from E. Finally the horizontal HG and the vertical FG are drawn. Fu will be taken as perpendicular to nu, and the small triangle FEu similar to the triangle HFG, and Eu equal to the line element FE; whence if one assumes that BL $= 1$ and DE $= x$; then MC $= 1 + L/l$; EC $= x + 1 + L/l$; HG $=$ CE $-$ BL $= x + L/l$; from these equalities

$$Fu = \left(\frac{1}{l} + \frac{x}{L}\right) \times FE$$

It remains that Hn be determined: It is to be observed that the thread LE, while it is being contracted, draws the body E upwards in the direction [LE]; while it draws the other body L back obliquely in its own direction Ln. Therefore with this said, Ln to Eu will be as DE to LE or, rather, as $x$ to $L$. But in addition Ln to Eu is, reciprocally, as the mass of the body L to the mass of the body E, that is, directly, as $M$ to $m$. When the proportions are combined, $Ln : Eu = Mx : mL$, whence, writing FE for Eu, Ln $= (Mx/mL)$FE; and since Hn $=$ HL $-$ Ln, it follows that

$$Hn = \left(\frac{1}{l} - \frac{Mx}{mL}\right) \times FE.$$

The accelerating forces on the bodies H and F are therefore as $1/l + x/L$ to $1/l - Mx/mL$: To obtain isochronism, these forces are assumed proportional to the distances described, LB and EC, or rather, suppose that

$$\left(\frac{1}{l} + \frac{x}{L}\right) : \left(\frac{1}{l} + \frac{Mx}{mL}\right) = 1 : \left(x + 1 + \frac{L}{l}\right),$$

and when this is solved, one obtains

$$x = \frac{mL - ml - ML - Ml \pm \sqrt{\{4mMLL + (ml + ML + Ml - mL)^2\}}}{2ML}$$

If MC or $1 + L/l$ is added to this, one obtains

$$CE = \frac{mL - ml + ML + Ml \pm \sqrt{\{4mMLL + (ml + ML + Ml - mL)^2\}}}{2ML} \times BL,$$

entirely as in the first part of this argument, third theorem, prop. 7.

    III. If the same things are assumed, the length of the simple isochronous pendulum will be equal to

$$\frac{2mLl}{mL + ml + Ml + ML \mp \sqrt{\{4mMLL + (ml + ML + Ml - mL)^2\}}},$$

the reason for which you understand from the fact that if a simple pendulum of length AH or $l$ were considered, the accelerating force in this simple pendulum would be to the accelerating force on the body H in our composite pendulum as H$l$ to H$n$, that is, as $1/l$ to $1/l - Mx/mL$: moreover, the lengths of the isochronous pendula are in the reciprocal ratio of the accelerating forces. Therefore the length of the desired pendulum to the length AH will be as $1/l$ to $1/l - Mx/mL$, whence is found the length of the isochronous pendulum $= mLl/(mL - Mlx)$, and assuming for $x$ the value found above, the length will be, as was said, equal to

$$\frac{2mLl}{mL + ml + Ml + ML \mp \sqrt{\{4mMLL + (ml + ML + Ml - mL)^2\}}}$$

IV. If the thread is loaded with three bodies at H, F, and G, making isochronous and extremely small vibrations, and the mass of the highest body is assumed $= m$; of the middle body $= M$ and of the lowest $= \mu$: the distances AH $= l$; HF $= L$; FG $= \lambda$: the distance of the body H from the vertical line AP $= 1$; the distance of the body F from the same vertical line $= s$; then

$$\{(MMl\lambda + M\mu l\lambda)ss + (mMl\lambda + m\mu lL - mML\lambda - MMl\lambda - MML\lambda$$
$$+ m\mu l\lambda - M\mu l\lambda - M\mu L\lambda)s - m\mu l\lambda - mMl\lambda\} \times \{(Ml\lambda + \mu l\lambda)s$$
$$- mL\lambda - Ml\lambda - ML\lambda - \mu l\lambda - \mu L\lambda + mlL\} = mm\mu llLLs.$$

The distance of the lowest body from the vertical line will be, for any root $s$, equal to

$$\left(\frac{MM\lambda}{m\mu L} + \frac{M\lambda}{mL}\right)ss + \left(1 + \frac{\lambda}{L} + \frac{M\lambda}{\mu L} - \frac{M\lambda}{\mu l} - \frac{MM\lambda}{m\mu L} - \frac{MM\lambda}{m\mu l} - \frac{M\lambda}{mL} - \frac{M\lambda}{ml}\right)s$$
$$- \frac{M\lambda}{\mu L} - \frac{\lambda}{L}.$$

In order to demonstrate these things, it is again assumed that the lowest thread FG is easily extended and that thus the body G, accelerated by the natural force of gravity, descends vertically from G to $s$ in a given infinitesimal interval of time while both the higher bodies are accelerated just as in the first figure by making their arcs H$n$ and F$u$. Moreover, it is clear that if G$s$ in the second figure is assumed equal to the descent FE in the first figure, the arcs H$n$ and F$u$ will be the same in both figures; therefore, by the preceding paragraph H$n = (1/l - Mx/mL) \times$ G$s$ and F$u = (1/l + x/L) \times$ G$s$, where $x$ is the infinitesimal interval u$M$ drawn perpendicular to the prolongation of A$n$, just as next we understand by $y$ the line element $yv$, which is perpendicular to the prolongation of $nu$. Now draw the horizontal FQ and the vertical QG, and having taken u$y =$ FG, draw G$y$. These things having been prepared for the calculation, it is now as in the previous case to be imagined that the straight line u$s$ is contracted to its original length FG: so the body will be raised from $s$ to $y$ or $r$ (however $yr$ is as nothing compared with G$y$); the higher bodies are again drawn

Fig. 2

back from n to o and from u to m: and thus it is clear that the accelerating force on each body in its natural direction will be to the natural force of gravitation as Ho, Fm and Gy to Gs, respectively. It remains therefore to find an expression for each of these elements, having observed that the arcs Hn, Fu etc. are nothing in comparison with the distances of the bodies from the vertical line. By doing the calculation correctly, one finds that $FL = \lambda/l + (\lambda/L)x + y$ and, since $FG : FQ = Gs : Gy$, one obtains

$$Gy = \left(\frac{1}{l} + \frac{x}{L} + \frac{y}{\lambda}\right) \times Gs.$$

Next one has to find the size of the retrogradations of the bodies (which, at this point, are supposed to be at u and n) which are made while the lowest body is raised from s to y or to r. It is to be observed that the tension contracting the thread us is carried equally in the other threads. Therefore again as in the above paragraph um to sy or rather to Gs will be in a ratio composed of vy to uy and of the mass $\mu$ to the mass $M$: Whence $um = (\mu y/M\lambda) \times Gs$, and when this is subtracted from Fu or rather from $(1/l + x/L) \times Gs$, there arises

$$Fm = \left(\frac{1}{l} + \frac{x}{L} - \frac{\mu y}{M\lambda}\right) \times Gs.$$

Finally, since the highest body is not affected by the shift of the middle body from u to m, there will be, as before no to ys or Gs in a ratio composed of Mu to un and of the mass $\mu$ to the mass $m$; whence $no = (\mu x/ML) \times Gs$. And subtracting this from nH or from $(1/l - Mx/mL) \times Gs$, one obtains

$$Ho = \left(\frac{1}{l} - \frac{Mx}{mL} - \frac{\mu x}{mL}\right) \times Gs.$$

Now that we have thus found the acceleration of all bodies in their true directions, we must make them proportional to their distances from the vertical line, yP, uC, and nB, or rather to the quantities $(1 + L/l + \lambda/l + x + (\lambda/L)x + y)$, $(1 + L/l + x)$ and (1), in order to obtain isochronism: Two equations determining the values $x$ and $y$ will be obtained: and if next it is assumed that $1 + L/l + x = s$, the same equation for $x$ will be found as that which we examined above and which we undertook to analyze.

V. The acceleration of the body H expressed by Ho or rather by $(1/l - Mx/mL - \mu x/mL) \times Gs$ is to the acceleration of the same body when the two lower bodies are absent, expressed by $(1/l) \times Gs$, as $1/l - Mx/mL - \mu x/mL$ to $1/l$ or rather as $mL - Mlx - \mu lx$ to $mL$. It follows thence that the length of the isochronous pendulum is

$$\frac{mLl}{mL - Mlx - \mu lx}$$

Moreover, that this does not differ from that which was given in the first part in the thirteenth proposition you will see if you replace the $x$ given there by $s$ or by $1 + L/l + x$, just as the definitions made by us require.

VI. Let there now be many bodies and as many as you wish, say B, C, D, E, F. Let [the line of] each thread be extended and let the sines of the angles BAN (AN is vertical), CBG, DCH, EDL, FEM be designated by $p$, $q$, $r$, $s$, $t$; let the masses of the bodies be designated by the same attached letters. I say that, with the natural force of gravity assumed $= 1$, the accelerating forces of the bodies according to their directions will be as follows.

Fig. 3

$$\text{on } B = p - \frac{C + D + E + F}{B}\, q,$$

$$\text{on } C = p + q - \frac{D + E + F}{C}\, r,$$

$$\text{on } D = p + q + r - \frac{E + F}{D}\, s,$$

$$\text{on } E = p + q + r + s - \frac{F}{E}\, t,$$

$$\text{on } F = p + q + r + s + t;$$

That this is the true law of the accelerating forces you will see if you suppose that a sixth body is attached by its own thread to the lowest body considered here, and if you next do the calculation as we did, in the case of three bodies in the fourth paragraph, by imagining, naturally, that the lowest body is accelerated downwards by the natural force of gravity, leaving the rest of the system to move according to its own nature, and then that the same body is elevated again by the contraction of the thread. Thus you will see that this law of acceleration, now accessible, is continued from five bodies to six, and thence to seven and thus to any number one pleases.

From the sines of the angles the distances of the bodies from the vertical line can be deduced; and if you make the distances proportional to the corresponding accelerating forces you will have as many equations as unknowns; thus everything desired can be correctly determined.

VII. Suppose now that the bodies are infinite in number and of equal mass, arranged with infinitesimal and equal distances between them: you will have the ideal uniform chain suspended from one end, such as AC or AF. In the [figure] the element Mm or Nn is to be considered infinitely small, with MN and mn drawn perpendicular to AC and mo drawn parallel to AC: It is assumed that Am or An $= s$ (these do not differ since they are infinitesimally separated); mM or nN $= ds$, taken as a constant element; the length of the whole chain AF $= l$; mn $= y$; Mo $= dy$. By the preceding

Fig. 4

theorem (assuming that the natural force of gravity $= 1$) the accelerating force on m will be equal to the sum of all the sines of the contact angles between A and m, diminished by the third-proportional of the corpuscle at m, the sum of all the corpuscles in MF, and the sine of the contact angle at M.

Thus the accelerating force on M is

$$\int \frac{ddy}{ds} - \frac{(l-s)\,ddy}{ds^2}.$$

Since one requires isochronism, the accelerating force is proportional to the ordinate MN; that is, with $n$ taken as a constant

$$\int \frac{ddy}{ds} - \frac{(l-s)\,ddy}{ds^2} = \frac{y}{n}.$$

The integral of the first term is taken without the addition of a constant, since nothing is to be added here. Thus

$$\frac{dy}{ds} - \frac{(l-s)\,ddy}{ds^2} = [+]\frac{y}{n}.$$

Finally suppose that $l-s$ or FM or CN $= x$, then $-dy/dx - x\,ddy/dx^2 = y/n$, or

$$n\,dy\,dx + nx\,ddy = -y\,dx^2,$$

an equation that indicates the nature of the curve AF: Since its integral is not evident, I assumed

$$y = \alpha - \beta x - \gamma xx - \delta x^3 - \varepsilon x^4 - \text{etc.}$$

$$dy = -\beta\,dx - 2\gamma x\,dx - 3\delta xx\,dx - 4\varepsilon x^3\,dx - \text{etc.}$$

$$ddy = -2\gamma\,dx^2 - 2\cdot 3\delta x\,dx^2 - 3\cdot 4\varepsilon xx\,dx^2 - \text{etc.}$$

When these values are substituted and next the equation divided by $dx^2$, there arises

$$-\beta - 2\gamma x - 3\delta xx + 4\varepsilon x^3 - \text{etc.}$$

$$-2\gamma x - 2\cdot 3\delta xx - 3\cdot 4\varepsilon x^3 - \text{etc.}$$

$$+\frac{\alpha}{n} - \frac{\beta}{n}x - \frac{\gamma}{n}xx - \frac{\delta}{n}x^3 - \text{etc.} = 0.$$

An equation that one satisfies by assuming $\alpha = 1$; $\beta = 1/n$; $\gamma = -1/4nn$; $\delta = 1/(4\cdot 9n^3)$; $\varepsilon = -1/(4\cdot 9\cdot 16n^4)$ etc. whence

$$y = 1 - \frac{x}{n} + \frac{xx}{4nn} - \frac{x^3}{4\cdot 9n^3} + \frac{x^4}{4\cdot 9\cdot 16n^4} - \text{etc.}$$

where 1 is to be understood as the distance of the lowest point from the vertical: and since at $x = l$ it has been supposed that $y = 0$, there will be also

$$1 - \frac{l}{n} + \frac{ll}{4nn} - \frac{l^3}{4 \cdot 9n^3} + \frac{l^4}{4 \cdot 9 \cdot 16n^4} - \text{etc.} = 0.$$

From this is to be derived the value of the letter $n$, which expresses the length of the subtangent at F. These things demonstrate the truth of the theorem which is the eighth in the preceding dissertation.

VIII. In order to obtain the length of the isochronous pendulum, the accelerating force at the point F has to be found; by par. VI. it will be equal to the sum of the sines of all the contact angles from A to F, thus, it is $\int [+] ddy/dx$ or $[+] dy/dx$, taken at $x = 0$; and whence $[+] dy/dx = -1/n$. And so the accelerating force at F to the natural accelerating force is as 1 to $n$. If a simple pendulum has length $l$, the accelerating force on it is $1/l$ at the same distance from the vertical line; therefore the accelerating force at the extremity of the chain is to the accelerating force of the simple pendulum of the same length as $l$ to $n$; and therefore the length of the simple pendulum isochronous with the chain is $n$, as in the ninth theorem of the previous dissertation.

IX. The tenth and eleventh theorems depend only on the addition of an appropriate constant, therefore let me not undertake their demonstrations here as this would be excessively easy: but the twelfth will be deduced, again from par. VI, in the following way:

The corpuscles are considered to be infinite in number and assumed to be at equal distances from each other but now of unequal weights. Thus one will have the ideal chain that is unequally thick. Let it be so arranged that $\xi$ corresponds to the length FM$(x)$, with $\xi$ designating some function of $x$. By par. VII there will be an accelerating force on M equal to $\int -ddy/dx - \xi \, ddy/d\xi \, dx = y/n$, or since $dx$ is constant, there will be $-dy/dx - \xi \, ddy/d\xi \, dx = y/n$, or $n \, d\xi \, dy + n\xi d \, dy = -y \, d\xi \, dx$, or finally

$$-\frac{n\xi \, dy}{dx} = \int y \, dx,$$

as the twelfth theorem shows, about which there was discussion. The demonstration will be made more intelligible if the seventh paragraph is considered at the same time.

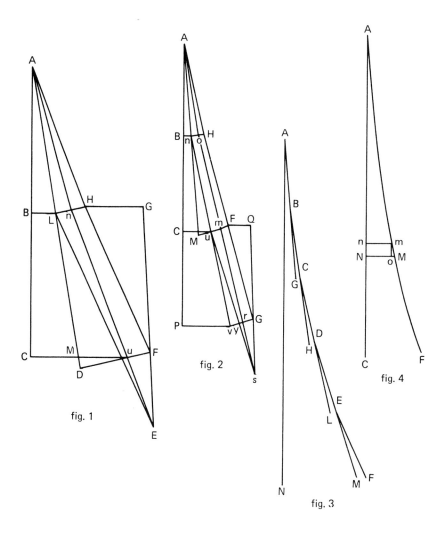

fig. 1

fig. 2

fig. 3

fig. 4

# Bibliography

Bernoulli, Daniel

[1] Methodus universalis determinandae curvaturae fili. *Comm. Acad. Sci. Petrop.*\* 3 [1728], 62–69 (1732).

[2] Observationes de seriebus recurrentibus. *Comm. Acad. Sci. Petrop.* 3 [1728], 85–100 (1732).

[3] Notationes de aequationibus, quae progrediuntur in infinitum, earumque resolutione per methodum serierum recurrentium: ut et de nova serierum specie. *Comm. Acad. Sci. Petrop.* 5 [1730–31], 63–82 (1738).

[4] Theoremata de oscillationibus corporum filo flexili connexorum et catenae verticaliter suspensae. *Comm. Acad. Sci. Petrop.* 6 [1732–33], 108–122 (1738).

[5] Demonstrationes theorematum suorum de oscillationibus corporum filo flexili connexorum et catenae verticaliter suspensae. *Comm. Acad. Sci. Petrop.* 7 [1734–35], 162–173 (1740).

[6] Commentationes de status aequilibrii corporum humido insidentium. *Comm. Acad. Sci. Petrop.* 10 [1738], 147–163 (1747).

[7] De motibus oscillatoriis corporum humido insidentium. *Comm. Acad. Sci. Petrop.* 11 [1739], 100–115 (1750).

[8] Commentationes de oscillationibus compositis praesertim iis quae fiunt in corporibus ex filo flexili suspensis. *Comm. Acad. Sci. Petrop.* 12 [1740], 97–108 (1750).

[9] De vibrationibus et sono laminarum elasticarum. *Comm. Acad. Sci. Petrop.* 13 [1741–43], 105–120 (1751).

[10] De sonis multifariis quos laminae elasticae diversimode edunt disquisitiones mechanico-geometricae experimentis acusticis illustratae et confirmatae. *Comm. Acad. Sci. Petrop.* 13 [1741–43], 167–196 (1751).

[11] Reflexions et eclaircissemens sur les nouvelles vibrations des cordes exposées dans les mémoires de l'Académie de 1747 & 1748. *Mém. Acad. Roy. Sci. Berlin* 9 [1753], 147–172 (1755).

[12] Recherches physiques, mecaniques et analytiques, sur le son & sur les tons des tuyaux d'orgues différemment construits. *Mém. Acad. Roy. Sci. Paris* [1762], 431–485 (Paris, 1764).

Bernoulli, Johann

[1] *Opera Omnia*, 4 vols. Lausanne & Geneva: Bousquet, 1742. Facsimile reprint Hildesheim: George Olms, 1968.

[2] Extrait d'une lettre . . . contenant l'application de sa règle du centre de balancement a toutes sortes de figures. *Mém. Acad. Roy. Sci. Paris* [1703], 78ff, 272–283 (2nd edition, Paris, 1720).

———

\* *Comm. Acad. Sci. Petrop.* = Commentarii academiae scientiarum imperialis Petropolitanae, Akademiia nauk S.S.S.R., Leningrad.

[3] Theoremata selecta pro conservatione virium vivarum demonstranda ... excerpta ex epistolis datis ad filium Danielem, 11 Oct. et 20 Dec. (stil. nov.) 1727. *Comm. Acad. Sci. Petrop.* 2 [1727], 200–207 (1729). Also in *Opera Omnia*, III, pp. 124–130.

[4] Meditationes de chordis vibrantibus. *Comm. Acad. Sci. Petrop.* 3 [1728], 13–28 (1732). Also in *Opera Omnia*, III, pp. 198–210.

Bernoulli, Johann II

[1] *Recherches physiques et géometriques sur la question: comment se fait la propagation de la lumiere.* Paris 1736. In: *Recueil des pieces qui ont remporté les prix de l'Académie Royale des Science* 3 (1752).

Cohen, I. Bernard

[1] *Introduction to Newton's 'Principia'.* Cambridge: Harvard University Press, 1971.

Cramer, Gabriel

[1] *Dissertatio physico-mathematica de sono*, Geneva, 1722.

[2] Extrait de quelques lettres de M. Cramer, ... écrites à M. de Mairan, ... et d'une Reponse de M. de Mairan à M. Cramer. *Journal des Sçavans* 124 [June 1741], 167–202 (Amsterdam, 1741).

D'Alembert, Jean Le Rond

[1] CORDES (vibrations des). In: *Encyclopédie, ou Dictionnaire Raisonné des Sciences, des Arts et des Metiers*, ed. D. Diderot & J. L. d'Alembert, VII, pp. 451–453. Bern & Lausanne, 1782.

Diderot, Denis

[1] Mémoires sur differents sujets de Mathematiques. In: *Oeuvres*, X, pp. 386–512. Paris: J. L. J. Briere, 1821.

Dostrovsky, Sigalia

[1] Early Vibration Theory: Physics and Music in the Seventeenth Century, *Arch. Hist. Exact Sci.* 14, 169–218 (1975).

Dostrovsky, Sigalia & Cannon, John T.

[1] Musikalische Akustik 1600–1750. In: *Geschichte der Musiktheorie.* Berlin: Staatliches Institut fur Musikforschung, to appear.

Ellis, A. J.

[1] H. Helmholtz, *On the Sensations of Tone*, trans. and ed. with notes by A. J. Ellis. 2nd edn. London, 1885. Facs. rep. New York: Dover 1954.

Eneström, G.

[1] Der Briefwechsel zwischen Leonhard Euler und Johann I Bernoulli. *Bibliotheca Mathematica*, ser. 3, IV, 344–388 (1903); V, 248–291 (1904); VI (1905), 16–87.

[2] Der Briefwechsel zwischen Leonhard Euler und Daniel Bernoulli. *Bibliotheca Mathematica*, ser. 3, VII [1906–07], 136–156.

Euler, Leonhard

[1] Notebook Hl. Euler Archive, University of Basel.

[2] De oscillationibus annulorum elasticorum. In: *Opera Omnia*, ser. 2, XI part 1, pp. 378–382. Laúsanne, 1957.

[3] *Dissertatio physica de sono.* Basel, 1727. Reprinted in *Opera Omnia*, ser. 3, I, pp. 182–196.

[4] Solutio problematis de invenienda curva quam format lamina utcunque elastica in singulis punctis a potentiis quibuscunque sollicitata. *Comm. Acad. Sci. Petrop.* 3 [1728], 70–84 (1732). Reprinted in *Opera Omnia*, ser. 2, X, 1–16.

[5] De minimis oscillationibus corporum tam rigidorum quam flexibilium methodus nova et facilis. *Comm. Acad. Sci. Petrop.* 7 [1734–5], 99–122 (1740). Reprinted in *Opera Omnia*, ser. 2, X, 17–34.

[6] De oscillationibus fili flexilis quotcunque pondusculis onusti. *Comm. Acad. Sci. Petrop.* 8 [1736], 30–47 (1741). Reprinted in *Opera Omnia*, ser. 2, X, pp. 35–49.

[7] *Tentamen novae theoriae musicae.* St. Petersburg, 1739. Reprinted in *Opera Omnia*, ser. 3, I, pp. 197–427. Facsimile reproduction New York: Broude Brothers. English translation by C. S. Smith (Ph.D. dissertation, Indiana University, 1960).

[8] De motu oscillatorio corporum flexibilium. *Comm. Acad. Sci. Petrop.* 13 [1741–43], 124–166 (1751). Reprinted in *Opera Omnia*, ser. 2, X, pp. 132–164.

[9] De curvis elasticis. In: *Methodus inveniendi lineas curve maximi minimive proprietate gaudentes*, Additamentum I. Lausanne & Geneva 1744. English translation by W. A. Oldfather, C. A. Ellis, & D. M. Brown, in *Isis* 20, 70–160 (1933).

[10] Judicium de libello Domini de la Croix. . . . C. Truesdell, ed. In *Opera Omnia*, ser. 2, XVIII, pp. 413–417 [*Scientia navalis*, Appendix I]. Zurich, 1967.

[11] Notae ad Responsiones viri illustris de la Croix. . . . C. Truesdell, ed. *Ibid.* pp. 418–427 [Appendix II].

Fellman, E. A.
[1] Jakob Hermann. *Dict. of Scientific Biography*, VI. New York: Charles Scribner's Sons, 1972.

Fuss, P. H.
[1] ed., *Correspondance Mathematique et Physique de Quelques Célèbres Géometres du XVIII ème Siècle.* 2 vols. St. Petersburg, 1843. Facs. rep. New York: Johnson, 1968.

Hall, A. Rupert & Trilling, Laura
[1] See Newton [2].

Hermann, Jakob
[1] *Phoronomia, sive de viribus et motibus corporum solidorum et fluidorum libri duo.* Amsterdam, 1716.

[2] De vibrationibus chordarum tensarum, *Acta Eruditorum* [1716], 370–377.

Huygens, Christiaan
[1] *Oeuvres Complètes.* 22 vols. The Hague: Nijhoff, 1888–1950.

Jeans, Susi
[1] Taylor. *The New Grove Dictionary of Music and Musicians.* 20 vols. Stanley Sadie, ed. London: Macmillan, 1980.

Kline, Morris
[1] *Mathematical Thought from Ancient to Modern Times.* New York; Oxford University Press, 1972.

Lagrange, Joseph Louis
[1] Recherches sur la nature et la propagation du son. *Miscellanea Taurinensia* 1 (1759). In: *Oeuvres de Lagrange*, I, Paris 1867.

Lana Terzi, Francesco de [Franciscus Tertius de Lanis]
[1] *Magisterii Naturae et Artis*, II. Brescia, 1686.

Laplace, Pierre Simon
[1] *Traité de mécanique céleste.* 5 vols. Paris, 1799–1825. Facs. rep. Brussels: Culture et Civilisation, 1967.

McGuire, J. E. & Rattansi, P. M.
[1] Newton and the 'Pipes of Pan,' *Notes and Records of the Royal Society of London* 21, 108–143 (1966).

Maclaurin, Colin
[1] *A Treatise on Fluxions.* 2 vols. Edinburgh, 1742.

Mairan, Jean Jaccques d'Ortus de
 [1] *Hist. Acad. Roy. Sci. Paris* 32 [1720], 14–15 (Amsterdam, 1724).
 [2] Discours sur la propagation du son dans les differens tons qui le modifient, *Mém. Acad. Roy. Sci. Paris* 65 [1737], 1–87 (Amsterdam, 1741).
Manuel, Frank
 [1] *A Portrait of Isaac Newton.* Cambridge: Harvard University Press, 1968.
Newton, Isaac
 [1] *Philosophiae Naturalis Principia Mathematica.* 1$^{st}$ edn. London, 1687; 2$^{nd}$ edn. Cambridge, 1713; 3$^{rd}$ edn. London, 1726. Translation of 3$^{rd}$ edn. by Andrew Motte, *The Mathematical Principles of Natural Philosophy.* London, 1729. Facs. rep. London: Dawson's, 1968.
 [2] *The Correspondence of Isaac Newton,* V, ed. A. Rupert Hall and Laura Trilling. Cambridge, 1975.
Rameau, Jean-Philippe
 [1] *Nouveau Système de musique theorique.* Paris, 1726. Facs. rep. New York: Broude Brothers, 1965.
Robartes, Francis
 [1] A discourse concerning the musical notes of the trumpet and the trumpet marine, and of defects of the same, *Philos. Trans. Roy. Soc. London* XVII, 559–563 (1692).
Robison, John
 [1] Trumpet Marine. In: *Encyclopedia Britannica.* 3$^{rd}$ edn. (supplement), pp. 399–403. Edinburgh, 1803. Also in: *A System of Natural Philosophy* IV, pp. 486–500. Edinburgh, 1822.
Sauveur, Joseph
 [1] Sur la determination d'un son fixe. *Hist. Acad. Roy. Sci. Paris* [1700], 131–140 (Paris, 1703).
 [2] Maniere de trouver le son fixe. In: Système general des intervalles des sons. *Mém. Acad. Roy. Sci. Paris* [1701], pp. 357–360 (Paris, 1704).
 [3] Sur les cordes sonores. . . . *Hist. Acad. Roy. Sci. Paris* [1713], 68–75 (Paris, 1716).
 [4] Rapport des sons des cordes d'instruments de musique aux flèches des cordes; et nouvelle determination des sons fixes. *Mém. Acad. Roy. Sci. Paris* [1713], 324–348 (Paris, 1716).
Smith, Robert
 [1] *Harmonics, Or the Philosophy of Musical Sounds.* Cambridge, 1749. Reprinted, with an introduction by J. Murray Barbour, New York: Da Capo Press, 1966.
Spiess, O.
 [1] ed. *Der Briefwechsel von Johann Bernoulli,* I. Basel, 1955.
Stoot, W. F.
 [1] Some aspects of naval architecture in the XVIII$^{th}$ century. *Trans. of the Royal Institution of Naval Architects* 101, 31–46 (1959).
Struik, D. J.
 [1] *A Source Book in Mathematics, 1200–1800.* Cambridge, Mass.: Harvard University Press, 1969.
Taylor, Brook
 [1] De motu nervi tensi, *Philos. Trans. Roy. Soc. London* 28 [1713], 26–32 (1714). Translation: Of the Motion of a tense String, *Philos. Trans. (abridged)* 6, 14–17.
 [2] *Methodus incrementorum directa et inversa.* 1$^{st}$ edn. London, 1715; 2$^{nd}$ edn. London, 1717.
 [3] *Contemplatio philosophica.* London, 1793.

Truesdell, Clifford A.
   [1] The Rational Mechanics of Flexible or Elastic Bodies, 1638–1788. In:
   *Leonhardi Euleri Opera Omnia*, ser. 2, XI part 2. Zurich, 1960.
   [2] The Theory of Aerial Sound, 1687–1788. In: *Leonhardi Euleri Opera
   Omnia*, ser. 2, XIII, pp. VII–CXVIII. Lausanne, 1955.
Watson, G. N.
   [1] *Theory of Bessel Functions*. Cambridge, 1944.
Westfall, R. S.
   [1] *Force in Newton's Physics*. London: MacDonald; New York: American
   Elsevier, 1971.
Whittaker, E.
   [1] *A History of the Theories of Aether and Electricity*. London 1951. Reprinted
   New York: Harper torchbooks 1960.
Wolf, Rudolf
   [1] *Biographien zur Kulturgeschichte der Schweiz*, III, Zurich, 1860.
Young, William
   [1] [Biography of Taylor]. In: Taylor [3], pp. 1–40.

# Index

# Studies in the History of Mathematics
# and Physical Sciences